Contributions to a Computer-Based Theory of Strategies

Nicholas V. Findler

Contributions to a
Computer-Based Theory
of Strategies

With 61 Figures

Springer-Verlag Berlin Heidelberg New York
London Paris Tokyo Hong Kong

Nicholas V. Findler
Computer Science Department
Arizona State University
Tempe, AZ 85287-5406, USA

658.403
F 49 C

CR Subject Classification (1987):
I.2.4, I.2.6, I.2.8, J (all chapters) and C.2.1, C.2.4, I.2.1 (Chap. 6),
D.2.6, I.2.7, I.3.2-3, I.3.6 (Chap. 7), G.1.2, G.1.6, G.3 (Chap. 5),
H.4.2, I.2.1, I.6.4 (Chap. 3), I.2.2, I.2.7 (Chap. 4)

ISBN 3-540-52634-X Springer-Verlag Berlin Heidelberg New York
ISBN 0-387-52634-X Springer-Verlag New York Berlin Heidelberg

Library of Congress Cataloging-in-Publication Data
Findler, N.V.   Contributions to a computer-based theory of strategies / Nicholas V.
Findler.   p.   cm.
  Includes bibliographical references (p.) and index.
  ISBN 0-387-52634-X (U.S.)
1. Decision-making – Data processing. 2. Strategic planning – Data processing.
I. Title.   T57.95.F55   1990   658.4'03'0285 – dc20   90-9857   CIP

© Springer-Verlag Berlin Heidelberg 1990
Printed in the United States of America

2145/3140-543210 – Printed on acid-free paper

Preface

It is customary to describe the motivation for writing another book — information explosion may threaten our intellectual survival 'almost' as much as the population explosion endangers the integrity of our physical existence. I hope the reader will be generous enough to accept the following.

For some ten years, between about 1970 and 1980, my students and I were interested in computer studies of human decision making under uncertainty and risk. Out of these investigations has emerged an area of inquiry that we, tentatively and among ourselves, have called *Theory of Strategies.* This name looked rather presumptuous to us and we just kept on working without 'publicizing' the catchy umbrella term for our projects. We have now completed a fair number of working systems and have applied them to several problem domains. It seems the right time to provide a systematic overview of this work, stop for a moment and see where we are going from now on.

I hope to have adopted the right style for a 'monograph in Artificial Intelligence' — as opposed to the style of articles in journals or conference proceedings. There is a growing number of such monographs in Artificial Intelligence, describing in some detail particular projects of interest and stimulating further efforts in the area. It would be gratifying to know that I have succeeded in joining their rank.

The book is aimed at researchers, university and college teachers, graduate and senior undergraduate students in Computer Science, Management Science and certain areas of engineering. The desirable prerequisites are a certain level of maturity in Computer Science and Mathematics, and some familiarity with the basic concepts of Artificial Intelligence. The subject matter may also be a part of a semester-long graduate seminar on, say, 'Advanced Topics in Artificial Intelligence'.

I hope that students and beginning researchers will get a glimpse of how research problems are identified and abstracted to a level at which they are both simple and still realistic enough; how they are divided into meaningful and relatively independent components which, when implemented and properly integrated, can be fine-tuned to yield effective and efficient problem solving systems. The experienced researcher may benefit by becoming familiar with the detailed mechanisms employed in several large-scale, complex programming systems that solve theoretically and practically important problems.

Similarly to most areas of study in Artificial Intelligence, our work employs concepts, methods and techniques originating also from other disciplines, such as Mathematics, Statistics, Logic, Operations Research, Linguistics and Psychology. The projects described have clearly defined research objectives which they attempt to accomplish. We make plausible and realistic assumptions about the environment in which the programming systems will work, and about the constraints and requirements they have to satisfy. The resulting systems are fairly general purpose as well as robust with regard to the assumptions made.

The applications of the different systems range over various aspects of air traffic control, automatic verification and validation of discrete-event simulation models, econometric model building, distributed planning systems for manufacturing, control of traffic lights, balancing of assembly line resources, and others. These applications are both interesting and important. The quality of the results is expected to be better than of those obtained via some traditional approach.

A few words are in order to describe the organization of the book. A brief introduction discusses how a *theory* may be "discovered" by a computer and embedded in its program — without getting involved in the underlying, controversial philosophical issues. The second chapter defines some basic concepts in reference to strategies, decision making and planning. Each of the next six chapters deal with one large-scale project: the Quasi-Optimizer, the Advice Taker/Inquirer, the Generalized Production Rule System, Distributed Planning and Problem Solving Systems and a Predictive Man-Machine Environment, respectively. Each of these chapters have a similar structure; namely, they present introductory concepts and research objectives, describe the approach, the more important issues of implementation and applications, a short summary, and acknowledgements of the efforts by coworkers. At the end of the book, an overall summary, references and a subject index, listing the main concepts of the book, are provided.

Last but not least, it is my pleasure to acknowledge my debt to my many past and present students for their conceptual and technical contributions.

First, I must single out (in an alphabetical order) Tim Bickmore, Cem Bozsahin, Bob Cromp, Ji Gao, Qing Ge, Ron Lo, Neal Mazur, and Uttam Sengupta. I am also grateful to many others, in particular to Mike Belofsky, Parvathy Bhaskaran, John Brown, Steve Feuerstein, Laurie Ihrig, Cher Lo, Ron Lo, João Martins, Ernesto Morgado, Bede McCall, George Sicherman, and Han You. From among several faculty collaborators, I note specially Jan van Leeuwen and Wai-Wan Tsang.

The work described in this book has been partially supported by the NSF Grant DCR-85-14363, AFOSR Grants 81-0220 and 82-0340, U.S. Department of Transportation Contract DTRS-5683-C-00001, U.S. Coast Guard Contracts DTCG-50-86-P-01074, DTCG-86-C-8042 and DTCG-88-C-80630, and the Digital Equipment Corporation External Research Grants 158 and 774.

I spent my sabbatical year 1988-89 at the Institute of Computer Science of the University of Zurich in Switzerland, where I did most of the writing of this book. I had access to wonderful computing facilities, and enjoyed a most congenial and friendly atmosphere. My sincere thanks are due to all colleagues and, in particular, to the Institute's director, Professor Kurt Bauknecht. Last but not least, I am grateful to Springer-Verlag for their help in bringing this book to reality.

Tempe, Arizona and Zurich, Switzerland, 1989   Nicholas V. Findler

Contents

List of Figures

List of Tables

1. Introduction

In addition to to the millions of 'mundane' tasks in business, engineering and scientific calculations, computers have also been used in discovering, modifying and verifying *scientific theories.* Let us discuss briefly and informally what is meant by a scientific theory and how can it be *computer-based.* This area is probably the most controversial in Artificial Intelligence and is closely related to the philosophical problem of how machines can manifest creativity and originality. Further, I note that the epistemological relation between *theory* and *program* is held by some people as (i) a possibly complete identity, or (ii) the appropriate program is considered as the model of a theory, or (iii) the program is taken as a heuristic tool in theory formation but with no formal connection to theory. I am not a philosopher and it is not my intention to cover such questions in this book. In forming one's opinion, the empirical observations are obviously strongly affected by attitude and perception. The following informal discussion should be regarded simply as a guide to orient the reader before embarking on the main topics.

Usually, the scientific community first[1] has empirical data. The objects and events, their properties and relations to each other — as expressed by the data — exhibit certain regularities that scientists want to explain in a systematic, concise and 'universally applicable' manner. ("In parsimony lies beauty" may be the basic tenet of our scientific education.) In doing that, scientists refer to things observed as well as to terms they posit as having some physical reality — although these terms may also originate from and be suggested by the scientist's philosophical or aesthetic judgement. It is realistic to refer to the *context* that guides and constrains the search for regularity in theory formation. (See, for example, [196].) Scientist have *a priori* models (spiced by expectations and even prejudices of a sort) which tell them which data to look for and, most importantly, how to interpret

---

[1]There are a few notable exceptions to having empirical data first. The most famous is Maxwell's equations about electromagnetic waves, which predated the corresponding experimental results of Hertz and others.

them — the "inductive leap" in drawing conclusions. The selected pattern is validated by the model since the given set of data may be matched by a large (infinite?) number of 'acceptable' patterns. Anderson [4] introduces a somewhat different set of ideas and makes distinction among the terms *frameworks*, *theories* and *models* as follows:

> A framework is a general pool of constructs for understanding a domain, but it is not tightly enough organized to constitute a predictive theory. However, it is possible to sample from this pool, tie the constructs together with additional details, and come up with a predictive theory. One might regard "information-processing psychology" as such a framework. ... One cannot evaluate a framework according to the standard verificational logic associated with scientific theories. That is, the production system framework makes no unique empirical prediction that distinguishes it from, say, a schema system framework. Rather, one judges a framework in terms of the success, or fruitfulness, of the theories it generates. ... A theory is a precise deductive system that is more general than a model. ... A model is the application of a theory to a specific phenomenon...

As Thomas Kuhn [116] so convincingly demonstrated, the proposing, accepting and validating of a model is an inherently social affair. The interactions between an individual scientist and the conservative control structure of the society of scientists result in changes of scientific theories, ranging from refinements and extensions to violent revolutions. (It seems to me less than realistic, however, to consider the evolution of science as being solely subject to such a set of laws and mechanisms, without recognizing the cognitive and even the affective factors in the behavior of individual scientists.)

Scientific theories, in general, are *judged* on the basis of how large their scope of validity and accuracy are in explaining observed phenomena and predicting unobserved ones, how simple ('elegant') they are, how well they fit in the prevailing structure of existing theories, how many and how plausible hypotheses they employ, and how flexible they are through their various admissible interpretations.

The formal-axiomatic expression of scientific theories, as exemplified by Euclidean geometry, classical mechanics or electromagnetic theory, may not be attainable by most branches of natural, life and social sciences. Theory construction in a particular scientific domain is constrained by the demands and possibilities imposed by the experiential data and the method of observing them. In some abstract sense, scientists in the 'non-exact' fields seek a compromise between the empirical basis of scientific knowledge, on one hand, and the systematic coherence and structure of scientific understanding, on the other.

We should not delve any further into the basic problems of scientific inquiry, the styles of explanatory procedures, the methods of stating and interpreting concepts and principles. We keep in mind that there is a justifiable diversity in the approaches to the question of how scientific theories are created, and we should strive, where possible, to extend and integrate them.

Next, we describe the concept of *computer-based scientific theories*, using three well-known families of projects as examples, and point out where our own work fits into this pattern.

Lenat, in a series of articles [131-135] described a methodology as to how heuristics can lead to the discovery of new knowledge. His approach was basically *empirical* — in the sense that the reasoning processes were guided and controlled by examples — as opposed to the *analytical* reasoning mode also common in mathematical discoveries. In the course of arriving at new knowledge, the number and the power of heuristics in Lenat's program change, and new representations become available — all under the guidance of heuristics. He provided his first program, *AM*, with an initially small set of basic operators and facts about number theory. Using some clever heuristics (for example, with reference to the assumed properties 'interesting' and 'important'), *AM* (re)discovered several important concepts in number theory, such as that of prime numbers and arithmetic functions. In Lenat's later program, *EURISKO*, the heuristics themselves are also subject to the same treatment as the concepts in *AM* were. The resulting processes of generalization, specialization and analogy can create new and modify existing heuristics. *EURISKO* thus became more powerful and versatile than *AM*.

Second, in contrast with above *theory-driven* approach, Langley, Bradshaw Simon and Zytow [21, 120-125] developed a family of programs, different versions of *BACON*, that discover empirical laws in a *data-driven* fashion. A small number of heuristics (about likely functional relations, common divisors and evaluating differences of different order between tabulated values of variables) detect patterns in numerical and nominal experimental data. Hypotheses on relations between variables are formed, tested, and accepted or rejected. The hypotheses — as if they were data — can then be recursively subjected to heuristics for further analysis. We should point out, as the authors also do, that in addition to the descriptive discovery (numerical pattern finding) process they are currently concerned with, there are other major tasks involved in scientific discovery, such as the decision as to which data should be examined, how to compensate for noise (measurement errors, the effect of unknown and unconsidered variables), how to formulate explicatory theories, how to use these for prediction, and so on. In other words, this group of researchers has ignored the role of models we discussed before and concentrated on representing error-free experimental data in a concise and simple manner. The question of acquiring, representing and organizing knowledge is central in all scientific paradigms, and the development of the related methodology serves as the gauge by which one can measure progress in the discipline.

The third family of projects are often denoted by the term *simulation of cognitive behavior*. Here we deal with work that represents an intersection between the interests in Artificial Intelligence and Cognitive Psychology. As such investigations started during the early history of Artificial Intelligence, the researcher observes the behavior of one or several humans in a particular task environment and makes a record of the major manifested aspects of behavior, such as spoken and written utterances. The corresponding record is called the 'protocol' of the experiment. This phase of the study is then followed by writing a program that, once put to the same task, acts in a manner similar or identical to the human. In other words, the trace of the program should reproduce in a reasonable manner the experimental protocol. (We must emphasize, though, that the goal is, of course, to reproduce the *right behavior* motivated by the *right reason*.) Unsatisfactory model behavior can be corrected by changing the underlying assumptions about the participating information structures and processes, by modifying assumed causal relations, parameters and the like. Having completed this activity, the researcher obtains a so-called *sufficiency theory*, that is a collection of mechanisms which is sufficient (but may not be uniquely necessary) to explain, understand and predict certain selected aspects of human thought processes under a variety of conditions. Such processes can be involved in problem solving, decision making, verbal learning, concept formation, and so on. (See, for example, Feigenbaum's *EPAM* project [55] or the comprehensive book by Newell and Simon on human problem solving [163]. There are several high-quality general descriptive books about AI and this area of study, such as [7] or [186].)

Current research in the above area has extended the scope of investigations. In addition to studying the problems of language acquisition, memory processes, pattern recognition, procedural learning and other complex cognitive activities, answers are sought to such fundamental questions as what the role of and interaction between declarative and procedural knowledge is, how data-driven and goal-oriented processes coexist, and the like.

The research activity described in this book should be seen in the context of these three families of approaches. Our philosophy and results relate, perhaps most closely, to the third, last family of projects — with one significant difference. In the area of simulation of cognitive behavior, the explanations obtained are at a relatively *micro level*[1]. Our goals are stated more in the *engineering vein* and at a *macro level*. We have specified a certain set of objectives that we wish to attain economically, reliably and effectively — regardless of how humans deal with the problems at hand. If we can learn some lesson from the human approach, we simply add it to our tool kit in a pragmatic way but we do not wish to *simulate* humans — in fact, if possible, we would like to surpass the quality of human performance. In other words, we try to combine the 'psychology oriented' and the 'engineering oriented' approaches in an experimental manner for satisfactory/optimum results. The programs encompass a theory of the activity, which can then be applied to a variety of tasks.

---

[1]See, for example, Newell and Simon's definition [164] of the *physical symbol system*. It references atomic symbols, expressions composed of such symbols, and basic, elementary processes acting upon them. The authors argue that intelligent behavior *can* arise out of the computational manipulation of such symbolic structures.

2. Basic Concepts of Strategies, Decision Making, and Planning

People use the word *strategy* in a variety of different contexts. Its original meaning, "the art of the general" in ancient Greek, refers to the conduct of warfare[1]. The term has later assumed connotations ranging from statesmanship and management of national policy to diplomacy and economic planning. Ever since John von Neumann and Oskar Morgenstern [162] showed the similarity between game-like problems in economics, sociology, psychology and politics, the concept of strategy has become pervasive also in the social sciences. We also talk about "problem solving strategies", "corporate strategy" in a large business enterprise, and so on — whenever a sequence of goal-oriented actions is based on large-scale and long-range planning.

We adopt this, latter type of interpretation of the concept 'strategy'. Basically, we consider strategy as a decision making mechanism which observes and evaluates its environment, and — in response to it — prescribes some action. We will then gradually expand this idea in several directions as follows.

A strategy relies on *long-term considerations* in which the consequences of its decisions may remain relevant to the final outcome. Normally, several strategies confront each other. They may be *competitive*, with mutually exclusive goals, in a *conflict situation* or else they can be *neutral* in following a set of rules and without trying to 'win' in antagonistic confrontations. (Nature — without stretching the concepts too far — is probably the best example of the latter type of decision making entity.) We note that, in contrast, *tactics* have short-term goals and consider only certain, spatially or temporally local, aspects of the environment. Tactics may be thought to be coordinated and invoked by a strategy.

A strategy includes the means of evaluating the adversaries' situations and actions, scheduling its tactics and, by making use of the feedback from

---

[1] See the classical theoretical works by Karl von Clausewitz [31] and Antoine-Henri de Jomini [104].

the environment, adapting itself to a new or changing situation. We call the latter type '*learning strategies*', as opposed to static ones. We are particularly interested in the various processes of self-adaptation in strategies. One possible paradigm for it is the case in which the strategy modifies its tactics, both in terms of their contents and their inter-relation. The strategy thus gives tactics their mission and seeks to reap their results.

One component of the strategy decides on *which* variables are relevant — at a given place and time — to characterize the environment, and on how to interpret these variables. (As we discuss it later in the book, we have considered both numerical and symbolic variables of different types as well as such variables that can characterize the *history* of the development leading to the current situation.) It should be pointed out that the picture of the environment, as perceived by the strategy, is unclear. The set of variables chosen to describe it may be incomplete, some values may be missing or obscured by noise. Noise can be caused by latent (unknown or just unconsidered) factors, measurement errors or deliberate obfuscation by the adversaries (cf., bluffing in Poker). There may also be biases and conflicts within an organization (for example, due to rivalry between divisions or personalities) that could distort the picture of the environment and produce erroneous descriptions of situations.

The *action* prescribed by the strategy in response to the environment may be a single-step one or a *sequence of action steps* — resulting from a long-term planning process involved. Similarly, in addition to a *scalar* action, we must also consider *vector* actions. The components of the latter need not be completely independent of (orthogonal to) each other. For example, an executive wants to respond to the current market situation and calls for budget adjustments in different divisions, transfer of personnel, and changes in the product profile — all at the same time. The variables describing actions can again be of numerical and symbolic types.

The *goal* of the confrontation can also be a scalar or a vector. As before, the components of the goal vector need not be independent of each other — in fact, in real life, there may be even conflicts and constraints between them. A large company, for example, may wish to maximize both its profit and market share (under certain legal and financial conditions and restrictions). These goal components are certainly inter-dependent.

In a probabilistic and partial sense, the strategies *control* the (physical or social) environment and cause changes in its state. A *quality measure* of such control is needed. One would like to observe and evaluate not only how well the overall goal has been accomplished at the end but, also, how close the goal is after a certain selected action step. It is, therefore, desirable to dissect the strategy into meaningful and distinct components, if possible, and to associate these with good or bad intermediate outcomes. In other words, we would like to *assign credit* (see [154]), a measure of merit, to different strategy components in different situations.

The mathematical theory of games [162] has given us a conceptual framework, a useful terminology but few practical techniques to solve large-scale, complex, real-life problems of conflict. The methodology of Artificial Intelligence coupled with decision theory, utility theory and operations research should, however, make important contributions to this area, to the *analysis and synthesis of strategies*. We have coined the term *digital gaming* — an activity combining model building, simulation and learning programs — for the method we have used to study decision making strategies in competitive situations of conflict. Projects aimed at this problem represent the subject matter of this book.

The remainder of this chapter describes some basic concepts of decision making and planning. *Decision making* represents a time-sequence of choices which constitute the basis or reason for certain action. We can view decision making as a response mechanism operating in a particular task environment and guided by a strategy. The consequences of a decision may in advance be understood completely, partially or not at all — and in a deterministic or probabilistic manner.

The goals of decision making may be positive (to attain certain objectives) or negative (to avoid certain events). These goals may be of the short-term or the long-term type. In the latter case particularly, some *planning* activity precedes the actual decision making. Planning contrasts and, in some sense, tries out alternative methods of attack in a course-grained way. Planning can thus be thought of applying a crude model to a task. The task description contains the initial and the desired situations, and the planner computes a sequence of (major) steps toward achieving the goal, that is reaching the desired situation.

Plans are, in general, continually refined and incrementally modified. A part of the plan must take care of the *resource allocation* among the three categories of activity: gathering more information if necessary, refining and modifying the plan if necessary, and making progress in carrying out the plan. The modification of the plan makes economic sense as long as its expected cost is less than the anticipated increase in the utility of the modified plan. This simple and plausible statement implies that the human (or the computer program) about to make the relevant decision has the following pieces of information available:

- measures on the cost and complexity of planning,
- reliability values of alternative plans,
- utilities of individual goal achievement,
- probabilities of achieving goals as the consequence of a given ordered sequence of actions,
- trade-offs among alternative subgoals, etc.

However, this kind of information is rarely available and, even if it is, its reliability is rather low. The difficulty of an analytical approach is further aggravated by changing environments and goals, the need for real-time operation, and the lack of heuristic guidance in working with *combinatorially explosive* search domains.

Finally, we note that in actual real-life situations, the conditions for "rational decision making" do not prevail. Namely, a *rational decision maker* is assumed to be fully aware of

- the set of alternatives open to choice,
- the relationship between the outcomes (pay-offs) and his choice,
- the preference ordering among outcomes.

He can, therefore, make a choice the consequence of which is at least as good as that of any other choice. In other words, he can make consistently optimum decisions.

Simon [208] has replaced the above, idealized and abstract actor with the "satisficing man[1]". This individual's environment is complicated, vague and uncertain. Instead of optimizing, he is content if his current level of

---

[1]The following caveat is given. Words, such as 'man' and 'he' are used in the generic sense throughout the whole book and can refer to either gender.

aspiration is met ("satisficed"). This model considers incomplete and inconsistent information, conflicts, limited information processing ability, limited perceptions and desires. He is, however, able to learn and to modify both the structure and the contents of his goals. Such "bounded rationality" is realistic at the level of both individual and organizational decision making. We have set our own goals to adhere to this perspective. We would be glad if the realism of our assumptions and results could approximate that of the above model.

3. The Quasi-Optimizer (QO) System

3.1. Introduction and Research Objectives

One usually distinguishes in science between *descriptive* and *normative* *theories*. The former tells us what an entity does under certain conditions and, possibly, why. Cognitive psychology, sociology and demography are examples of the disciplines that provide descriptive aspects of human behavior. In contrast, normative theories tell us what to do in order to accomplish certain goals. Various optimization techniques and methods of Operations Research yield normative theories of well-defined and mathematically formulable tasks.

The descriptive aspects of decision making are investigated in a number of areas within psychology, such as social judgement theory, information integration theory and attribution theory. The normative aspects of decision making, on the other hand, are emphasized by mathematical decision theory, behavioral decision theory and psychological decision theory.

Decision making in modern environments has become significantly more difficult for humans for two reasons:

- time-constrained, critical choices have to be made from many alternatives that may also have uncertain consequences;
- the amount of information that has to be received and processed often surpasses human data handling and memory capabilities.

It is, therefore, essential to provide decision makers with techniques for organizing, analyzing and presenting information. We must increase our understanding of judgmental processes, develop models and procedures that depict human decision making, and establish techniques that can generate optimum/satisfactory decisions under all relevant conditions.

A recently emerging area in Artificial Intelligence produces knowledge-based *expert systems* (see, for example, [40, 98, 219]) that can also be considered decision support tools aiding the human decision maker. A significant and difficult problem faces the builders of such systems — how to acquire, represent in the computer, verify and optimize the large amount and different types of human expert knowledge needed. Several reports [39, 49, 159, 183] have described direct, interactive techniques. There are also user-friendly programming systems [23, 54, 83, 149] that prompt the expert or the "knowledge engineer", acting as an interpreter between the expert and the system, for production rules (*condition* —> *action* associations) [179]. Also, it should be necessary to reconcile the possibly different recommendations provided by several qualified human experts [184] in a satisfactory way (or even the same expert may well provide varying information about the same problem at different times).

After the above explanations, we discuss the research objectives of the *Quasi-Optimizer* (QO), and put them in a real-life, applications-oriented context [57, 70]. The paradigm is of rather general validity.

- We have wanted to find an efficient and effective computer representation of both static and dynamically changing strategies of decision making. Such representation will then have to be used for analytical studies either in a realistic simulated world or in direct interaction with the real world [58].
- We have wanted to develop a program that would observe a strategy in action, infer the rules underlying its decisions, and construct (induce) a descriptive theory — a computer model — of the strategy [57, 75].
- This program would identify meaningfully distinct components of a strategy and evaluate their effectiveness [78]. It would then select the most satisfactory components from different strategies and combine them in a 'super strategy[1]'. The latter would inherently contain some inconsistencies, redundancies and may also be incomplete. After these

---

[1] There are three reasons for which the quasi-optimum strategy is not (quite) a normative theory:

(i) The quasi-optimum strategy is at least as good as only any of the *input* strategies, from which the best components were selected. A strategy outside this set may employ certain controllers and indicators that are superior to those in the "training set" but have not been considered.

(ii) The quasi-optimum strategy is "normative" only in the statistical sense. Fluctuations, whether accidental or deliberate, in the adversary strategies may impair its performance, and it would have to adjust to such and *learning* adversary strategies, too. (Note, however, that a normative strategy of, say, Poker would also be the best only in the statistical sense. It would not win every single game because of the probabilistic factors involved in the hands received and in the seating arrangements. We would be satisfied if it wins more money than any of its opponents during an evening of play.)

(iii) The construction of the computer models of the input strategies and of the quasi-optimum strategy is based on approximate and fallible measurements.

deficiencies have been eliminated by a program, the resulting strategy will be called the *quasi-optimum* strategy [59, 61].
· We wish to make a step toward the automatic knowledge acquisition of certain types of human expertise. The computer model generated of the decision making strategy of a *human expert*, if constructed and verified appropriately, may serve as the *knowledge base* of an expert system. Further, the generation of the quasi-optimum strategy is in fact equivalent to the *fusing* of several experts' knowledge and *optimizing* it — an important and basic concern with expert systems, as noted before, for which no effective technique has been devised.

Let us now look at a real-life environment, and discuss how the behavior of the *QO* corresponds to certain events in it. Suppose there is a competitive market in which several organizations want to achieve some (at least partially) exclusive goals. Each organization perceives the environment by observing and measuring a set of variables *it* considers relevant. (Some exemplary variables include the real or assumed actions of the adversaries; the perceived state of the competition; the availability, extent, criticality and vulnerability of own resources; estimates of threats from others; etc.) Parts of the strategy of an organization is aimed at *interpreting and evaluating* these measurements, determining a *course of action* leading to goal attainment, and preventing the other competitors from attaining it. At any given moment, the 'rules' of the competition — including certain legal, financial and physical constraints — as well as the past and current actions of the participants determine the following state of the environment.

As we have said in the previous chapter, the picture of the environment is unclear because some information may be unavailable, missing[1] or obscured by noise. Noise can be caused by unknown causal factors of limited importance or intentional perturbations. If a competing organization's decisions are based on such incomplete or faulty information, its resources will be wasted and goal attainment will be further removed.

If a new organization wants to enter the competition, it must first develop a strategy for itself. This strategy would preferably incorporate all the good aspects of the extant competitors' strategies. The acquisition of such knowledge is done through a period of observation. One can follow one of two possible approaches in this regard, at least in some cases. In the *passive*

---

[1] It is usual to distinguish between *risky* and *uncertain* types of information. In the former case, one knows the relevant *a priori* probability distributions of variables, occurrence of events, causal relations, and the like. In the latter case, such knowledge is not available.

mode of observation, the organization does not interfere with the environment but makes a record of the situations and actions as they happen to occur in an indefinitely long sequence of confrontations. (Note that in many instances, this is not only the natural but the only way of performing this task when there are no control variables accessible by the organization through which it could 'set up' the environment.) There are also useful by-products of the passive mode of observation, such as obtaining automatically the probability distribution of situations, the total ranges of situational variables, and those combinations of subranges of values of these variables that are excluded for some historical, legal or physical reasons. However, this method is inefficient and the observation may take a very long time before statistically significant results are obtained over the *whole* decision space.

The *active mode of observation* can be followed when the organization is able to set each relevant situational variable to any feasible value. (We have referred to this approach as '*experimentation under laboratory conditions*'.) A selected set of situations, that is certain combinations of values of situational variables, created are then presented to the decision making strategy under observation for a response to be made. We have called a situation and the associated response an *experiment*. The sequence of situations must be well designed, according to an appropriate statistical methodology. Several possible techniques in this category will be discussed in detail later. Let us just note that the active mode of observation is, in general, more effective than the passive one and therefore preferable.

The new organization now has a *model* of each strategy adhered to by an organization currently competing. It seems plausible to assume that no single organization does *everything* the best possible way. Each segment of the whole activity (e.g., product research and development, inventory and production control, responsiveness to market changes, personnel selection and assignment, marketing, budget allocation, etc.) may be done best by several *different* organizations. It would be advisable to adopt these best components and to merge them into a single organizational policy, a sort of super decision making strategy. There may, however, be some problems with the super strategy. Conflicts and logical inconsistencies can exists between certain components. Some tasks may be assigned to more than one division while some others are not taken care of by any. A systematic testing process must discover such mistakes and eliminate them. It is clear that this corrected strategy is better than, or at least as good as the best of, all originally observed strategy. We call this the *quasi-optimum strategy* for reasons described before.

We will discuss other types of actual applications of the *QO* later. The actions of the system in these are similar but the stated goals are rather different. This fact will support the contention that the *QO*, its design and implementation are fairly general, and can be customized to a variety of purposes.

3.2. A Brief Survey of Related Work

This section presents a few relevant contributions in the areas of decision trees and knowledge acquisition, as a prelude to the discussion of the QO system.

Decision tree is the major information structure employed by the QO system, as described in the next section. Early work in decision trees focused on creating optimal ones in terms of memory utilization by rearranging the nodes in the tree (using exponential time algorithms). Payne and Meisel [169] considered this optimization for binary trees. Martelli and Montanari [143] transformed decision tables into decision trees. Pagall [167] used decision trees to represent logical formulae in disjunctive normal form which allowed the efficient computation of their truth values.

Green [92] originated the concept of path entropy — a measure of search efficiency on a decision tree, which can be considered to be the cost of reaching a particular leaf. Findler and van Leeuwen [70] created a more global measure based on the "level complexity". Quinlan [181] established a method of constructing a decision tree given a set of examples. He used multi-valued categorical variables in the construction task. Quinlan's ID3 program [180] builds a decision tree iteratively from a subset (window) of all examples. The constructed tree is tested for all examples. If the test fails for some examples, the window is enlarged and the process is repeated.

Automated knowledge acquisition programs have been designed to elicit knowledge from an expert, usually in the form of *if-then* production rules [95]. Early expert systems — for example, Slagle's symbolic integration program SAINT [209], the formula manipulating system MACSYMA [144] and the HEARSAY-II program [52] used for natural language understanding — had all their knowledge hand-coded into the program, which was then difficult to debug and update.

Davis [40] developed a powerful knowledge acquisition tool, TEIRESIAS to work in conjunction with the MYCIN expert system [205]. The latter was designed to diagnose and to suggest treatment for infectious blood diseases. TEIRESIAS allows a user to enter rules utilizing, through keywords, a small but effective subset of natural language. A major feature of TEIRESIAS is that it also contains knowledge about what it knows (meta-knowledge). This feature allows it to guide the user in completing its knowledge base and to query the user about unknown terms.

Politakis' SEEK system [176] also works in the area of medical diagnosis but its knowledge acquisition consists of a more structured data entry. The expert creates data tables and rule tables that specify symptoms and test results leading to a particular diagnosis. SEEK tests the rule base against known cases and suggests changes to improve its knowledge. Noble [166] developed a knowledge acquisition system based on schemas. A schema gives a stereotyped description of a sequence of events. By utilizing schemas during knowledge acquisition, the system can emulate the expert after eliciting the necessary knowledge in an efficient manner. The KRITON system by Diederich [46] utilizes several methods, such as various interview techniques, protocol analysis and semantic text analysis, to glean information from an expert. Wilkins' ODYSSEUS [225] works similarly to an apprentice. While it "watches" an expert performing, the system is given an explanation of the reasoning behind the actions. Meta-Dendral [25, 26, 138] an expert system used for analyzing chemical compounds, enhances its knowledge base by automatically generating new rules on the basis of case analysis. Note also [53] and [110].

Expert system shells contain no domain knowledge in their programs. They supply the user with a rule structure, a mechanism for entering rules and an inference engine for processing rules. All domain knowledge is entered by the user in the rules created [9]. The EMYCIN program [150] was designed to create expert systems using the MYCIN framework. It provides the user with a mechanism for specifying general *if-then* rules and is most appropriate for deductive problems, such as medical diagnosis. KAS by Duda, Gaschnig and Hart [49] is also a knowledge acquisition tool which requires a knowledge engineer. It allows the user to create *if-then* rules, each with appropriate rule strengths. The knowledge is expressed in semantic networks.

Weiss and Kulikowski's EXPERT [221] contains *if-then* rules but these are not unrelated. The *then*-part of a rule may cause other rules to be activated or change some attribute (such as confidence) in other rules. EXPERT is used mostly to create diagnostic models in medicine and uses forward chaining to reach a diagnosis from the given data. OPS5 [83] supplies the programmer with a mechanism to create *if-then* rules and objects with associated attributes. Its control mechanism consists of searching the rulebase for all rules whose antecedents are satisfied and a conflict resolution scheme to select one to be activated. The programmer can also specify temporal relationships such as task ordering. Rosie [54] is another rule-based expert system language and was designed to be close to English both syntactically and semantically. The computer system configurer R1 [147] required significant direct effort by knowledge engineers.

3.3. The Approach

Let us first discuss the computer representation of a static strategy [57, 70]. The information structure chosen is the *decision tree* (see Fig. 3.3.1). Each *level* of the decision tree is associated with a *decision variable* — in our case a situational variable. (We will use the terms 'decision variable' and 'situational variable' interchangeably.) Starting from the top (root level), there are as many *branches* emanating from a node as the number of *subranges* of the decision variable of that level. (The number of branches is called the *out-degree* or *branching factor* of the node in question.) A subrange is defined by the criterion that no matter which value the decision variable assumes within it, the action remains the same. The *actions* or *strategy responses* are "attached" to the branches emanating from the last level nodes — they hang from the *leaves*. A path from the root down to a given leaf corresponds to a situation and its associated strategy response. In view of the above definition of a subrange, a pathway corresponds to a whole *subrange* of situations and one single action.

Next, we have to define the *maximum meaningful resolution* (*MMR*). It represents the smallest observable difference between two adjacent values of a decision variable, due to the imprecision inherent in measurements, irreproducibility of experiments, and the limited sensitivity of the relevant part of the environment. In other words, we cannot distinguish between two values within *MMR* for any practical purposes. This concept will be useful

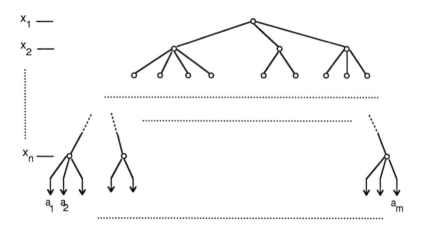

Fig. 3.3.1. — Schematic representation of a static decision tree. Here x_1, x_2, ... x_n are the decision variables and a_1, a_2, ... a_m are the actions

when the QO gradually builds up the model of a given strategy under observation. Initially, the subranges being established in the model have the width of MMR, which is then widened when the experiments warrant it.

A decision variable belongs to one of four types. It can be

- a *numerically oriented variable* if it assumes a "regular" number, a rank number or an ordinal numeral as a value;
- an *ordered categorical variable* if it assumes symbolic values that have an implied meaningful order, such as military ranks or "A-students", "B-students", ..., and can be mapped onto a numerical scale;
- an *unordered categorical variable* if the symbolic values assumed have no meaningful order, such as people's hair color or the continent they were born in;
- a *cyclical categorical variable* if the symbolic values assumed occur in a cyclically ordered manner, such as the days of the week.

Note that 'structural variables' (hierarchies, relations, priorities) belong to one of the types of categorical variables. Color is usually an unordered categorical variable unless, for example, the corresponding subrange of wave lengths matters. Furthermore, it is also possible to establish some well-chosen (computable) variables that characterize the crucial aspects of the *history of development* of the environment. (The response may depend on such variables, too, if the action sequence prescribed by the strategy adheres to a plan generated earlier.)

Whenever the strategy response is not a scalar but a vector action, the QO system has to establish a separate decision tree for each vector component. (The whole organization and the structure of these trees can, of course, be completely different. That is, the number and type of decision variables, the out-degrees of the nodes, and the subrange boundary points may all differ.)

A changing strategy will have a changing model. In order to deal with this problem in an effective manner, the QO has to distinguish between a convergent, learning strategy and one that fluctuates randomly. (The latter could also be the intention of the decision maker for good reasons, as the mathematical theory of games has shown us in prescribing randomly mixed strategies.) If we want the QO to generate a quasi-optimum strategy and the 'input strategies' include also a learning strategy, then only its best, final (asymptotic) version should be utilized. Therefore, the QO will have to extrapolate to the asymptotic form from a necessary number of successive models, each depicting the same strategy in the process of improvement at different stages [77].

We postpone the discussion of several other issues because they can be better understood when the different system modules are explained.

3.4. System Modules

There are six distinct but interacting modules in the QO (see Fig. 3.4.1). In addition, there are many utility and man-machine dialog programs which we will only touch upon when it is necessary for the general explanation. Furthermore, we have developed a very efficient two-way conversion program between decision tree and rule-base representations. The program optimizes the target representation. For example, when the resulting representation is of the rule-base type, all rules with identical right hand sides (same strategy response) are pooled together into one single rule. The Boolean function standing for the condition part on the left hand side is optimized by transforming it into a canonical expression whose evaluation is 'least expensive'. Similarly, decision trees resulting from the conversion from a

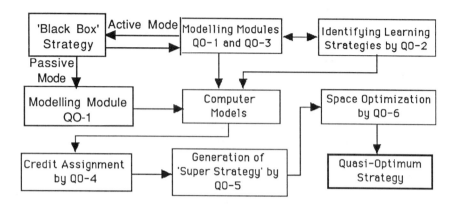

Fig. 3.4.1. — The relationship between the six major modules of the QO system

rule-base are subjected to the processes of memory requirement minimization in the *QO-6* module (see later).

3.4.1. *The QO-1 Module*

In an initial interaction with the system by responding to its prompted questions, the user specifies which are the 'likely' decision variables, what is their type, total range, *MMR*, which are the action variables, their type, and total range. The *QO-1* then starts gathering information and gradually builds up a decision tree [75]. It identifies the *relevant* decision variables and ignores the others in the superset given by the user. This happens in a conservative manner, according to an algorithm that computes a numerical *measure of relevancy*, between zero and one, for each given decision variable. This measure depends on how much change in the variable causes a 'unit' change in the strategy response on the average, across the total range of the decision variable. The user can, accordingly and in an iterative manner, select the best set of decision variables. Such action is particularly important when the danger of combinatorial explosion threatens.

If the user decides that the *QO-1* will follow the *passive mode of observation*, it is necessary for him to have at least an approximate idea as to whether there have been enough experiments carried out at a certain time point or still more are needed. Both numerical and graphical aids have been created for this decision [76].

As a numerical aid, the user can request the ratio between the number of experiments performed up to that point and the '*cardinality*' of the decision space, *C.* The latter is defined as the product of the cardinalities of the individual decision variables,

$$C = \prod_{(i)} c_i$$

When the decision variable is numerically oriented, its cardinality, c_i equals the ratio of its total range, T_i and its *MMR*, m_i

$$c_i = T_i / m_i$$

The cardinality of a categorical variable is equal to the number of values it can assume.

The user can also get the ratio of 'repeated' experiments and the total number of experiments so far. (A repeated experiment is one in which the value of each decision variable falls into the same *block* as at least once before. Blocks — arbitrary subranges of values — for each variable must be specified by the user at the beginning if repeated experiments are of concern. In a more general way, the user can specify a subset of decision variables, and ask for the ratio of repeated and total number of experiments in any combination of blocks.) A high ratio normally indicates that the model built is of lesser credibility.

Different graphical aids are also available to the user. He can obtain histograms of individual decision variable values. There is an important correction necessary here in order to provide for the user an unbiased image of the situation and the full benefit of the graphical aid. We call the correction the *normalization of frequency distributions* with respect to excluded regions. We have noted before that values of decision variables in certain combinations of subranges may not occur for some historical, legal or physical reason. Consider Figs. 3.4.2a and 3.4.2b.

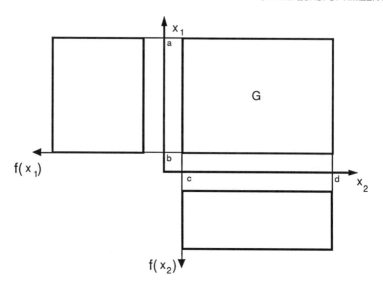

Fig. 3.4.2a. — Decision space **G** of two variables x_1 and x_2, both of uniform probability distribution; there are no excluded regions

Let us assume for the sake of simplicity that there are only two decision variables, x_1 and x_2, both uniformly distributed and with total ranges (a, b) and (c, d), respectively. If there are no excluded regions in the decision space G, as in Fig. 3.4.2a, the histograms obtained after a sufficient number of experiments should approximate the solid lines for $f(x_1)$ and $f(x_2)$. However, the excluded regions g_1 and g_2, as shown in Fig. 3.4.2b, would cause indentations in the histograms. Such an 'unjustified' graphical output could confuse the user who may think that not enough experiments have yet been performed to even out 'statistical accidents'. To avoid such problems, the system must compensate for the lacking datapoints in the excluded regions and *normalize* the histograms in such cases.

One more comment is due about a certain advantage of the passive mode of observation. It is often preferable to the active mode of observation with human subjects who may not react consistently enough — according to their 'deterministic' strategy — if the experimenter presents them with an (ostensibly arbitrary) sequence of environments to respond to.

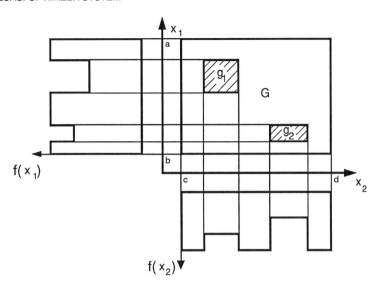

Fig. 3.4.2b. — Same as in Fig. 3.4.2a except there are two excluded regions, g_1 and g_2. The experimentally obtained histograms have to be normalized regarding the excluded regions, which would then yield a uniform distribution

If the user decides on the *active mode of observation*, he has two distinct choices. The first one relies on a statistical design that is prepared by the system *before* the experiments begin. There are two further options here. The user can have either an *exhaustive* experimentation or the equivalent of *binary chopping*. The former is very expensive since the method usually prescribes a very large number of experiments, equal to the cardinality of the decision space — which is not feasible in most practical cases.

In the binary chopping mode, *QO-1* assumes that the hypersurface representing the strategy responses in the multi-dimensional decision space is a weakly monotonic function of every decision variable. (In other words, the curve in any planar section of this hypersurface must not 'fluctuate' but will either continuously increase/stagnate or decrease/stagnate; that is, the partial derivatives of the response are everywhere either semi-positive or semi-negative.) As a first step, the two end values of every decision variable's total range, T_i are selected for experiments and the corresponding strategy responses obtained. If the difference between the two responses is greater than a threshold value (the desired level of precision), ΔR, then another experiment is performed at the midpoint of the total range. Further and further midpoints of subranges are selected (cf. binary chopping) as long as the difference in response values at the two endpoints of the previous subrange is still greater than ΔR — see Fig. 3.4.3.

There is an important *inductive discovery* process that the user of *QO-1* (and *QO-3* described below) can employ. These modules will correlate certain probabilistic components of the strategy response with identifiable regions in the decision space. Its significance is as follows.

The strategy response may have two types of probabilistic components:

·   relatively small ones (due, for example, to measurement errors), which we overcome by repeated experimentation, and
·   relatively large ones (due, for example, to unknown or unconsidered variables having 'localized' effects in certain regions of the decision space) that produce extreme values, so-called statistical '*outliers*'.

Another discovery process is described next. The system identifies the outliers and, if meaningful, computes the boundaries of the decision space within which a particular type and size-range of outliers fall with greater than a user-specified level of probability. (This is shown, in the case of three decision variables, on Fig. 3.4.4 in a schematic way.) If the 'volume' of this region is small, it indicates that there is a significant causal relationship between the decision variables involved and the probabilistic component of

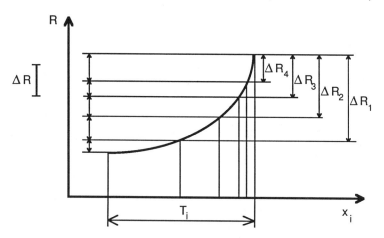

Fig. 3.4.3. — For the explanation of the binary chopping technique. Note that at the end of iteration, $\Delta R_4 < \Delta R$

the strategy response. We note that humans are prone to act according to such principle — human "randomized" behavior is strongly environment-dependent. For example, in Poker games, the frequency and intensity of bluffing are strongly correlated with the current state and the history of the game.

The second approach to the active mode of observation employs a *dynamically evolving* statistical design to be described with the *QO-3* module [66].

3.4.2. *The QO-2 Module*

The *QO-1* (and the *QO-3*) can model only static strategies. The *QO-2* [77] extends this facility to deal also with dynamically changing strategies. In general, strategies may vary randomly, exhibit some periodic behavior, or learn from experience till they reach a level of performance that is optimum within the limitations of their design and with reference to their adversaries in the 'training set'.

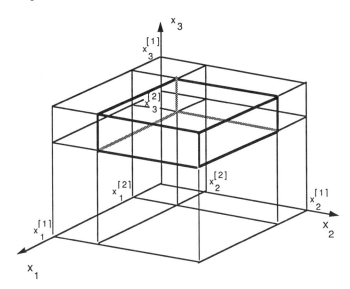

Fig. 3.4.4. — The result of an inductive discovery process. A probabilistic (scattered) response is correlated with a certain, computable region of the decision space. Extreme values fall, say, 95% of the time within the polyhedron marked with heavy lines

QO-2 is presented with a finite sequence of static models, decision trees, of a changing strategy. These models are essentially snapshots of the evolving strategy, taken when the 'learning component' is temporarily disabled. (*QO-2* invokes *QO-1* to generate the models a certain number of times — which is estimated by a statistical computation to be sufficient.) The result of the computation by *QO-2* is one of the following three possibilities:

- The learning process is *verified to be convergent* and the asymptotic form of the strategy is produced as the extrapolation of the sequence of decision trees. (This computation is rather complex since decision trees are multi-dimensional entities. The type and number of the relevant variables, the out-degrees of the nodes, the boundary points between the subranges, the type and number of the actions may all change in the course of a legitimate learning process.)
- The learning process *appears to be convergent* but there are not enough decision trees, snapshots, in the sequence for the desired level of statistical significance. *QO-2*, using certain plausible assumptions, calculates the approximate number of *additional* trees needed and gets them by invoking *QO-1* (or *QO-3*).
- There is no evidence for a convergent sequence — the strategy varies randomly. An "average" strategy can be computed if so desired.

Whenever a convergent learning strategy is identified, it is its asymptotic form that participates in the "training set" on which the computation of the quasi-optimum strategy is based.

3.4.3. *The QO-3 Module*

This module also provides an active mode of observation. Recall that *QO-1* offers the choice between the exhaustive and the binary chopping technique — both designs are made up *before* the experimentation starts. *QO-3* employs a *dynamically evolving design* [66] where each design depends on the results obtained so far. The experimental set-up created by *QO-3* is passed on to *QO-1*, which then asks the strategy to respond to.

The assumption about the monotonicity of the response surface is no longer necessary. The *total number* of experiments is minimized with the constraint that a user-defined precision is attained. It means that the response values at any two adjacent experimental points in the decision space never differ by more than the desired precision level — in other words, the response surface approximates a 'uniform level of sensitivity' (cf. ΔR in Fig. 3.4.3).

Putting the above idea in simple terms, the faster the response values change in a given region, the more experiments are needed there. Under ideal conditions, the difference in response values at two adjacent experimental points is constant. This would always be true if the response surface is a hyperplane and the experiments are performed according to a 'balanced incomplete block design' (see, e.g., [32, 103]). The experiments controlled by QO-3, in fact, do start with such a design. (Scientists like to model Nature as if it were linear...) The grid size is then continually refined *locally* whenever the behavior of the response surface warrants it. Eventually, QO-3 creates a grid of experimental points that closely satisfies the above criterion — the difference between the response values at two adjacent points is almost constant.

Finally, it should be pointed out that such an approach to statistically designed experiments can be of significant economic importance — the system specifies the parameters of a minimum number of experiments while a required level of precision in the experimental results is maintained. The experiments, of course, may well be expensive, dangerous or urgent.

3.4.4. *The QO-4 Module*

This subsystem [78] performs the *credit assignment* [154], a classical outstanding problem in Artificial Intelligence outlined in Chapter 2. The two major objectives are:

- to identify and distinguish strategy components;
- to associate with each component a quality measure of the outcome of the sequences of strategy actions to which the component contributes.

Note that the above agenda does not concern itself with what a 'strategy component' is or how the 'quality of the outcome' is measured. Let us consider a strategy S described by a certain decision tree. Let the environment be described by the *situation vector* $s(x_1, ..., x_n)$ each component of which is a value of the different situational variables. The *actions* prescribed by the strategy, $a_1, ..., a_m$, are attached to the leaf level of the decision tree. A given action may appear at several different leaves, which fact means that the strategy maps a certain *set of situations* into the same action: $S(s_j) => a_i$, $S(s_k) => a_i, \ldots$

Let us now assume that we can establish a *quality measure* of the consequences of an action in a situation, $q(s, a_i)$. This measure is, in the first approximation, independent of the strategy, and is a function of only the situation and the action. (A further refinement would also consider the long-range plans of the strategy at hand and evaluate the outcome in terms of also long-range consequences. In such a case, the *same responses* in the *same situation*, but prescribed by *different strategies* could possibly have *different quality measures*. The quality measure would depend also on how well the responses contribute to the long-term plans of the respective strategies.)

We attach the quality measures, on a scale between 0 and 100, next to the corresponding actions on the leaves of the decision tree, as shown on Fig. 3.4.5. (We consider a scalar value now; if the strategy response is a vector, the quality measure has to be a vector, too.) There will be two sliding boundary points, B and G, on this scale, denoting the separation points between "bad" and "neutral", and "neutral" and "good" ranges, respectively. Their final position will be determined by a learning process, as described below. The quality of an action's consequences will be considered "bad" if it is between 0 and B, and "good" if it is between G and 100 (see Fig. 3.4.6.).

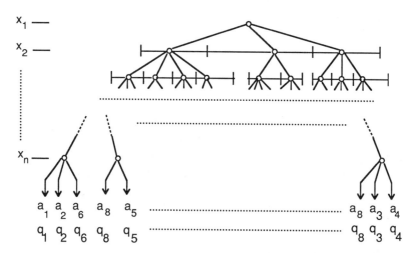

Fig. 3.4.5. — A schematic decision tree for the explanation of the credit assignment problem. The quality measures, q_i, of the consequences of each action, a_i, are also attached to the leaves

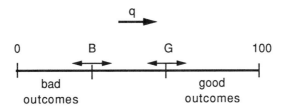

Fig. 3.4.6. — The scale of the quality of consequences. Common and discriminating features in sets of pathways on the decision tree help in defining strategy components

We propose the following operational definition of a strategy component. Recall that a certain action, a_i, may be prescribed by the strategy in a number of different situations. Let all the pathways that lead to the same action a_i form the class

$$U_i : \{u_j, u_k, ...\}.$$

Here the pathways $u_j, u_k, ...$ correspond to the respective situation vectors $s_j, s_k, ...$ Let us now define two subclasses of U_i — the subclasses of pathways that produce bad and good consequences, $U_i(b)$ and $U_i(g)$, respectively. We have, for example,

$$U_i(b) : \{u_m, u_n, ...\} \text{ and } U_i(g) : \{u_s, u_t, ...\}.$$

A *strategy component* is defined as the set of characteristic features discriminating $U_i(b)$ and $U_i(g)$. The *characteristic feature* of a pathway is the Boolean AND (in the general case, also the OR and NOT) of its atomic properties. *Discriminating* are those characteristic features that are not shared by the two sets of pathways and can, therefore, be used to contrast them. Finally, an *atomic property* of a pathway is the subrange of a decision variable value through which the pathway goes, at any level between the root and the leaves.

The algorithm developed for the credit assignment problem first forms *all* the characteristic features of the two subclasses of pathways $U_i(b)$ and $U_i(g)$. It then discards all but the discriminating features which are then considered to *represent the strategy component* associated with one of the actions, a_i.

We have also implemented a high-level learning process that maximizes the *power of discrimination* between the two subclasses. The location of the boundary points *B* and *G* is systematically changed on the quality scale and eventually optimized so that the sum of the probabilities of making Type I and Type II errors is minimum. These error types are analogous to those in statistical hypothesis testing, and refer to 'accepting a wrong pathway in' and 'rejecting a correct pathway from' its proper subclass of pathways, respectively.

The higher the power of discrimination (that is, the smaller the probability of making Type I and Type II errors), the more valuable that strategy component is. For a given situation, the *best action* is prescribed by that strategy (out of all considered) whose component associated with the situation has the best quality measure and the highest power of discrimination.

Finally, we have also studied different types of uncertainty and noise in the environment. These can lead to the following types of errors in the credit assignment task:

- a situation is not described exactly because of measurement errors, unknown or unconsidered variables;
- an action prescribed by the strategy in a given situation is not executed exactly because the environment has in the meanwhile changed;
- a certain action does not always have the same consequence in the situation assumed to be the same, for all the above reasons.

We have also developed fairly robust techniques that can deal with the above problems.

3.4.5. *The QO-5 Module*

This subsystem constructs the 'Super Strategy'. Using the results of *QO-4,* the best action is selected for each situation (subdomain). These are then combined so that the whole decision space (each possible situation) is responded to. (We note that the "best" actions may be further improved when the quasi-optimum strategy is put into operation to confront other strategies unconsidered for the 'training set'. Such a phase of activity has been termed 'experientialization' in the literature.)

3.4.6. *The QO-6 Module*

This module has the task of eliminating inconsistencies and redundancies from the Super Strategy while maintaining its completeness, effectiveness and soundness [63]. ('Soundness' refers to the relation between the optimality of the strategy component and of the whole strategy.)

The method of eliminating *inconsistencies* is straightforward and assures the maintaining of *soundness*, too. It is based on comparing and selecting the best components whenever several responses have been collected in overlapping subdomains of the decision space.

There can be several types of *redundancies*. We note three here:

- Identical or 'similar' subtrees hang from two adjacent branches emanating from a node. We have devised an efficient algorithm [81] that computes a measure of similarity between trees or subtrees and, when appropriate, merges the subtrees. (We note that non-adjacent branches can also be made adjacent by some non-linear transformation on the decision variable in question. We have, however, decided not to do this because the transformed decision variable would no longer be observable, it may have no physical meaning and its *MMR* would become irrelevant.)
- Some decision variables have little or no effect on the strategy response. Our calculation of the 'measure of relevancy' mentioned before is used to eliminate those variables whose relevancy is below a threshold value.
- Two decision variables can be combined into one, for example, when they are highly correlated. (The situations occurring in this case are such that a value of one of their descriptors implies a small possible subrange for the value of another descriptor.) We did not implement such simplification, beyond the measure used for less than relevant variables, for the reason stated in regard to the first type of redundancy.

3.5.   The Implementation

There were three basic requirements that we wanted to satisfy in implementing the system. *QO* had to be (fairly) general-purpose, flexible and interactive. Let us now discuss these requirements.

A *general-purpose system* is able to deal with a variety of task domains — with known limits and limitations. We will show in the next section two very different applications of the *QO*. They will shed light on how general the idea of a 'decision making mechanism' is and on the power associated with the methodology. Memory and computing time requirements are the usual limitations in the solution of large, realistic problems in AI. The quantitative concept of '*relevance*', computable for decision variables, enables the user to diminish the danger of combinatorial explosion. He can identify through an iterative process the most relevant *set* of decision variables. (Note that decision variables can be correlated with each other, and the relevance measure of a particular one depends also on which *other* decision variables participate in a model, at a given stage of the iteration. Therefore, the process of selecting the most relevant decision variables can be considered hill-climbing over a rather irregular surface.)

We have tried to enhance the system's *flexibility* by providing for it two modes of operation:

- if appropriate, *QO* can be "hooked up" electronically with the environment it is to model;
- a human can play the role of the 'I/O interface' for the *QO*. In other words, if the manner of observing the environment is not conducive to a direct tie with the computer, the user performs the information transfer in both directions.

In either of the above cases, it is necessary for the *QO* to acquire a lot of knowledge about the environment and this is done interactively.

A user-friendly *dialog system* makes the interaction with the system easy and error free. The dialog system is menu-driven and can operate in one of two modes. A *novice user* is led through the various menus at different levels and given instructions as to what needs to be entered at each step. The system makes sure that the user does not forget to define any system parameters needed and it offers him brief descriptions of items when so requested. The definition of the parameters takes place in a set order and their values are checked when possible. A *more experienced user* may want to define the running parameters in an arbitrary order or may load some or all of them from a file. A free-format dialog system enables him to do so. At the end, the user can view, edit, add to, delete from or output to a file the definitions made.

In either mode, each menu and prompting question has a *help* message associated with it, which can be obtained by entering a question mark. Each prompting question and the corresponding operation can be aborted by entering an exclamation point. All inputs are checked as to whether they are the right type (integer, real, literal...). The inputs are also checked for consistency ("Has it been defined before?" "Is it given within the right limits?" etc.) and completeness ("Has all necessary information been provided?"). There is even an automatic check of and suggested correction for spelling errors.

3.6.   Applications

Before we describe two major applications, we discuss a very important general area in which the QO system can be of quite some use. The acquisition, verification and appropriate storage of knowledge for expert systems have proved to be a slow and error-prone task; in fact, they are often referred to as the bottleneck in developing expert systems. There has been no satisfactory method to handle incomplete, changing and/or inconsistent knowledge obtained, nor a technique to deal with the sometimes conflicting views of different experts. The QO system makes some important steps toward the solution of these issues in tasks in which the human expert is faced with a structured selection or classification problem. The selection is to be made from a number of choices on the basis of available criteria — for example, in situation assessment, diagnosis, investment decisions, and the like. The QO system in these cases maps the expert's knowledge onto a flexible structure in an inductive manner, verifies the completeness and consistency of the knowledge, optimizes its representation (cf., [119], discovers the relative importance of the decision making criteria, and can itself become a machine expert. Finally, when several different human experts have been modelled by the QO system, it can reconcile their differences and construct a decision making mechanism that is at least as good or better than any single human expert analyzed.

The first area to which QO was applied is related to the predictive man-machine environment (see Chapter 8) that integrates the individual projects described in this book. A rather sophisticated 'Simulated Air Traffic Control (SATC) Environment' was created to contribute to the ultimate automation of various aspects of Air Traffic Control (ATC), an intellectually and economically important area of activity [62]. With reference to the QO, we have been interested in how the training and evaluation of air traffic controllers could be automated [60].

An *ATC trainee* is seated in front of a graphics terminal displaying either continuously changing scenarios of a near-airport situation or judiciously chosen static snapshots of the same. He is asked to make "real-time" decisions in either case. The *QO* system makes a computer model of his control strategy, and makes sure that the model is complete, unambiguous and consistent (within a *tolerated range*, which allowance is due to the non-deterministic environment and human response patterns). The consequences of the trainee's decisions are evaluated according to certain objective measures (the number of 'accidents' and 'near-accidents', the number and severity of violations of the vertical and horizontal plane separations as prescribed by statutory rules, the number of times certain physical limitations of aircraft capabilities have been exceeded, the number and extent of deviations from established flight paths and flight schedules, excess fuel consumption above the expected one, the number of instructions given for a certain objective to be accomplished, etc.). The trainee's strengths and shortcomings are reported back to him, and his degree of improvement over time is recorded, as well. Alternatively, ATC *experts* could be used in a set of experiments — an activity we have not pursued. The *QO* would first generate a computer model of each and then combine the best components of these into a quasi-optimum strategy of ATC.

As a by-product of these investigations, a *meta-strategy* may also be arrived at to indicate in which region of the whole decision space a given strategy is the most proficient — leading to an optimum way of dividing the total task into performing components or of assigning given individuals to subtasks.

The other application of *QO* aims at *verifying and validating discrete-event simulation models* [76]. It may be of interest to describe briefly the background of this project.

There are many real-life problems the analytical solution of which is impossible, difficult or prohibitively expensive. Computer-based simulation models (SMs) yield an inexpensive, flexible and transparent method of dealing with them. It is also true that it is often impossible to identify the consequences of certain simplifying assumptions inevitably used in the analytical approach. On the other hand, a family of simulation models, each of which based on somewhat different simplifying assumptions, should provide an answer to this problem.

No matter how carefully, however, an SM is constructed, various types of error are likely to occur in large, complex systems. These error types can be the following:

- the SM does not reproduce all situations that may appear in real life;
- the SM reproduces situations that are not allowed in real life (combinations of situational variables that are unacceptable for physical, legal or historical reasons);
- the 'responses' of the SM (actions, events) to given situations are not consistent (the stochastic components of the responses are outside acceptable levels and assume extreme values, so-called statistical 'outliers');
- the responses of the SM are identical or much too similar over a large domain of the situation space — a result which contradicts world knowledge and expectations.

The techniques of verifying the adequacy, accuracy and precision of SMs has so far been based on two techniques. First, one can perform *regression* or *time series analyses* on masses of output data elicited, to a large degree, indiscriminately from the model — an insensitive and uninformative method. Second, one can carry out "*structured walkthroughs*" — a non-rigorous and error-prone approach according to which the creator of the simulation program gives a narrative description of it to other people and receives their criticism.

There are several shortcomings inherent in the usual SM architecture. Namely, the user is unable to query the system as to what would happen after such and such *history* and in such and such a *situation*. A given SM appears to be a black box with limited and unchangeable input/output capabilities. The usual tendency of even computer-oriented simulation practitioners to believe in "what the computer says" underlines the need for a methodical and precise tool to verify and validate SMs. It would, of course, be a significant additional value if one could generate, preferably automatically, a higher-level *meta-model* of an SM that would not only serve as the tool for verification and validation but would also be capable of yielding the results of the SM in question. Furthermore, a highly desirable feature would enable one not only the construct new SMs according to some standard principles but also to modify relatively inexpensively *existing* SMs to accomplish the goals of verification and validation.

An SM can be considered as receiving both deterministic and pseudo-random input (most often generated by the program, itself) from which a *situation* is created. The SM then "responds" to it by computing some *action* or *event*. A component of this can be fed back to the SM to generate the next situation. Situations, actions and events can also be recorded on a *history file*, a selected part of which, possibly after some statistical processing, is output for the user (see Fig. 3.6.1).

The first objective is to generate automatically a meta-model (MM) of the SM, which has such characteristics that are conducive to the processes of verification and validation. The program generating the MM interacts with the user and interrogates him about certain fundamental properties of the SM and the real-life environment the model depicts. It then hooks up with the SM electronically through an interface to collect the information needed for the MM. A structural and statistical analysis of the MM reveals problems existing with the SM and these are communicated back to the user (see Fig. 3.6.2).

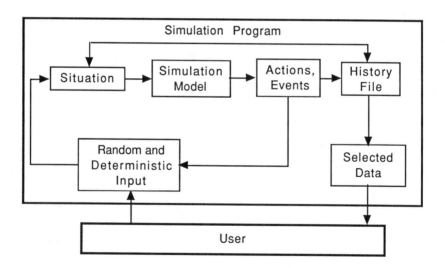

Fig. 3.6.1. — Schema of simulation programs

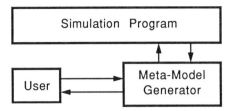

Fig. 3.6.2. — Schema for verifying and validating simulation models

If the results of SM *verification* are correct, that is none of the four types of errors listed above occur, and the SM appears "realistic" to the user (for example, it incorporates all the relevant situational variables), the user can *validate* the model. It can also be the case that it is less expensive in terms of machine time and memory to run the MM with an appropriately constructed control structure.

The *QO* system is very appropriate to perform the above operations. It considers the SM a 'decision making entity' and, after a dialog phase with the user, builds a model (in fact, a *meta-model*) thereof in the form of a decision tree — an information structure that lends itself to the verification process. The *QO* and the SM are electronically connected through an interface for information exchange. Any one of the options in the passive and active modes of observation can be employed. The properties of the tree are checked, with reference to the information given by the user, for completeness, consistency, excluded regions, legal chronological sequence of responses and other aspects of realism.

We have also created a control structure which enables the user to run the MM in lieu of the SM itself. There are many tools available to the user to obtain on-line textual and graphics help in running the system. He can generate even a *best composite meta-model* out of several meta-models of a *family of simulation models*. Here each simulation model is based on somewhat different assumptions about situational variables and their relation to the SM's response. This composite MM is then the least expensive and the most realistic to run.

In accordance with the technique described, we have both modified existing models and written new ones, in SIMSCRIPT, FORTRAN and Pascal.

Errors of different types were then deliberately introduced into the models. The system identified the errors, which were subsequently "corrected" in an iterative manner. After having verified the non-existence of the different types of errors, we were able to validate the models.

3.7. Summary

The *QO* project aims at satisfying the need for (fairly) domain-independent, automatic knowledge acquisition programs that can obtain, verify, fuse and optimize human expertise.

The *QO* system is capable of generating computer models (*descriptive theories*) of human decision making strategies. It can do it in two ways. In the *passive mode of observation*, the system does not or cannot interfere with the environment and records the characteristic features of the situations and the corresponding strategy responses to them. In the *active mode of observation*, the system designs a sequence of environments ('experiments') for the decision making strategy to respond to. The design of the experiments can be fixed in advance or can follow a dynamically evolving pattern that minimizes the total number of experiments needed for a user-specified level of precision.

A module of *QO* can ascertain whether a non-static strategy is in fact *learning*, that is converging to an asymptotic form, to which the program can then extrapolate. Another module can assign a quality measure ('credit') to the different strategy components identified by it, on the basis of their short- or long-term benefits in attaining certain goals.

The *QO* system can select the best components of several strategies and combine them in a 'Super Strategy'. The inconsistencies, incompletenesses and redundancies inherent in such a Super Strategy are eliminated and a Quasi-Optimum strategy is generated. Such a strategy is better, in the statistical sense, than any one participating in the 'training set'. The Quasi-Optimum strategy, therefore, corresponds to a *normative theory* within the limitations of the available information.

We have described two rather different areas of application — a possible approach to the automation of air traffic controllers' training and evaluation, and the automatic verification and validation of discrete-event simulation models.

Finally, it should be noted that there are several programs — some marketed as a commercial product, some being the result of scientific investigations (see, for example, [151, 181, 200]) — which *induce* decision trees or other information structures characterizing human expert behavior. We feel that the objectives reached, and the flexibility and power of our system exceed those of the other projects.

3.8.   Acknowledgements

The following collaborators with the *QO* project deserve my thanks: first of all Neal Mazur, then in alphabetical order Mike Belofsky, Tim Bickmore, Bob Cromp, Prof. Jan van Leeuwen, João Martins, Bede McCall, and George Sicherman.

4. The Advice Taker/Inquirer (AT/I)

4.1. Introduction and Research Objectives

The *Advice Taker/Inquirer* is a domain-independent program that is used to construct, fine-tune and oversee the operation of an expert system. It consists of two phases: a *learning phase*, during which a human expert teaches the system interactively about a domain in terms of principles, high-level examples, rules, and facts; and an *operational phase*, during which the program monitors the resulting expert system as it is applied to a domain of interest, and continually attempts to improve its performance by hypothesizing new rules and reorganizing existing knowledge.

First, let us briefly consider the history of the concepts involved with AT/I. The origin of the long-standing goal of AI, *automatic programming*[1], can probably be traced to a speculative paper by John McCarthy [146] on an Advice Taker system. His aim was to change the usual instruction-oriented mode of programming to a more human-like communication with machines. In the latter case, mostly declarative sentences describe the situations to be dealt with. The advantages include that

- the available knowledge can be easily utilized,
- the logical consequences of the declarative statements can be relied upon,
- the order of the declarations is usually of no importance,
- the programmer need not know what the effect of the individual declarations is.

(In contrast, imperative sentences can be performed more effectively and they need not depend on a priori knowledge.)

There have been two major lines of development of the original idea. A formal, theorem-proving approach, as exemplified by the deductive question-

---

[1]The term 'automatic programming' referenced the facilities provided by high-level computer languages during the early times of programming. In our context and in a more modern sense, it refers to a (desirable) system that can accept the specification of a problem in a very high-level language (possibly English), find the appropriate representation of data, generate the correct program and run it, and finally present the results to the user again in a high-level language.

answering systems of Black [13] and Green [91], finally culminated in logic programming [33, 114, 115]. In a similar vein, a knowledge-based approach is exemplified by [8, 29, 106, 187, 220]. A general overview is given in [12].

A less formal, heuristically-oriented approach has shown promise in game-playing programs. The advice-taking Chess computer of Zobrist and Carlson [231] received heuristic knowledge from an expert in a specially devised Chess language, such as ideas about beneficial attack patterns (e.g., "bring pawns to a forking attack position") and sound principles (e.g., " do not leave bishops and knights in the back row"). The resulting program can be debugged and modified faster and more easily than conventional programs. The domain-specific Advice Taker/Inquirer by Findler, Sicherman and Feuerstein [80] can be taught different Draw Poker strategies. The human expert provides *principles of play* (a Boolean combination of subranges of user-defined variables are associated with suggested general guidelines for actions) and *high-level examples* (sharply defined game situations are associated with direct game actions). *User-assigned player types* (simply called a certain name by the advisor) and *user-defined player types* (whose recent history has satisfied certain statistical criteria) can jointly or separately benefit from a set of given advices. Finally, the system is also able to discover when the advice received is vague, incomplete or contradictory, in which cases it makes inquiries and asks for clarification.

Mostow's program FOO [158] has a large number of rules which enable the expert's advice to be transformed into executable procedures (using also the expert's on-line guidance). Hayes-Roth, Klahr and Mostow [97] have identified five major steps that such a system in general must follow in transforming advice to a performing program:

- the system must *request* advice from an expert in a well-formulated way and inform him when the advice is contradictory and incomplete;
- the system must *interpret* and *represent* the advice internally;
- the advice (and the underlying processes) must be *operationalized* so that it can be applied to the relevant situations;
- the advice must be integrated into a knowledge base in a form that is testably neither too specific (which can potentially lead to conflicts) nor too general (which would indicate redundancy);
- finally, the advice must be empirically *evaluated* — if necessary, through a sequence of iterative improvements — by the collaborative effort of the human expert and the program so that the latter performs as expected.

From the point of view of knowledge elicitation/acquisition, the Advice Taker idea is conceptually related to a large number of projects which aim at inferring rules from empirical data (e.g., the Meta-Dendral [156]) and at building domain-specific knowledge bases (e.g., EMYCIN [149] and TEIRESIAS [40]). In TEIRESIAS, a restricted subset of natural language, coupled with keyword matching, was used to communicate expert advice. The program knows about its own structure and reasoning abilities, and forms expectations about the advice prior to parsing it. Such meta-knowledge contributes to the disambiguation of user input.

We present a few relevant definitions before the research objectives are discussed.

- A *principle* is a production rule-like statement that connects a Boolean combination of subranges of certain (both numerically and symbolically oriented) variables with some general action or guidelines for action.
- An *example* is a specific version of a principle — a sharply defined situation is associated with a direct, well-defined action. (The system will subsequently try to generalize one or several examples into a principle, possibly with the collaboration of the advisor.)
- An *assertion* is similar to an example but also contains a *certainty factor*, ranging between -1 and +1, which expresses the advisor's belief in the strength, quality or probability of the association between the situation and the recommended action (-1 corresponds to complete disbelief and +1 to complete belief).
- A *rule*[1] is similar to a principle but is not strategy-dependent. In other words, it represents a general relation that is valid in the environment at a given point or period of time.
- A *fact* is also a strategy-independent description of an object, action, event, relation, and their properties.
- A *definition* provides an explanation or circumscription of an entity (noun, verb, adjective, adverb, procedure, action sequence) in terms of elementary, system-resident entities, entities already defined, constants, parameters, statistical variables, Boolean and relational operators.
- An *order* or *instruction* given by the user to the system overrides the strategy imparted so far on one single occasion (for example, for the purposes of debugging the program).
- An *inquiry* can be given by the user to the system concerning principles, examples, definitions, hypotheses formed, exemplary actions in given

---

[1]Such 'rules' are to be distinguished from 'production rules', which will always be spelled out fully.

situations, values of statistical variables, etc. The system can also make an inquiry and ask the advisor when it finds the advice given so far vague or in conflict with previously provided advice.

- An *entity type* can be either *named* by the advisor as a category to which he would then assign entities or it can be *defined* by the advisor. (See the user-assigned and user-defined player types in [80] referenced above.) In a defined entity type, the categories are specified in terms of Boolean combinations of subranges of statistical variables. The system regularly tests and computes which entity would belong to which category over a given time period.

- An advice is considered *inconsistent* with a previously given one if the two recommend different actions in an identical situation. Note that the two specifications of the situation (the 'IF part' of the production rules in question) do not have to be identical — one can subsume the other. In other words, the condition part of one advice may be more general than that of the other.

- An advice is *redundant* if it is found to be identical with another advice already given. Again, this case can arise if the two advices lead to identical action in the relevant *decision space* only and not necessarily in *all possible situations*.

- Advices generate an *incomplete* knowledge base if there are situations in the relevant decision space for which no action is recommended.

- Finally, we will use the terms *advisor* and *expert* interchangeably in reference to the human source of the knowledge provided. In contrast, the *user* — the same or a different person — only employs the system for a certain task.

The two major schemas of *knowledge representation* used are *frames* [155] and *decision trees* [70]. Frames allow for the efficient storage and accessing of knowledge about the definition of objects in the environment. Decision trees are utilized for compilation and analysis of the expert's procedural advice. By combining the power of these two representations, the system can alert the advisor to the above named problems in the imparted knowledge.

The *research objectives* are as follows [36]:

- The domain-independent AT/I should be able to accept, verify and store domain-specific strategical *advice* of human experts; form on that basis a *decision making strategy*; *apply* it to given tasks; *evaluate* and *improve* the level of performance of the strategy, possibly in interaction with human users.

- The *communication* between the system and the advisor or users should be in a high-level language, preferably a subset of English, but also enhanced with menu-driven options in responding to prompted questions. AT/I must be able to acquire new terms, both declarative and procedural in nature, from the advisor's vocabulary.

- AT/I should be able to *detect*, and *attempt* on its own to resolve, *ambiguities* and apparent *conflicts* in the advice provided. If its syntactic and semantic analyses fail to produce a unique interpretation of the advisor's input, the system would query the advisor in a manner that should lead to the proper interpretation (*well-formed inquiries*). The detection of the discrepancies — incompleteness, inconsistency and redundancy — may take place either in the *learning phase*, when the necessary knowledge is being acquired, or in the *operational phase*, when the resulting strategy, in fact an expert system, is applied to system-generated test cases or user-provided tasks.

- The system must recognize *undefined terms* in an advice being given and remind the advisor to take care of such.

- AT/I should be able to *assess* a task and determine if it has the necessary knowledge and resources to solve the task.

- *Causal hypotheses* between goals and apparently successful decisions should be inductively generated by AT/I and added to its knowledge base.

- When necessary to make inquiries of the advisor, the system should be able to construct the proper *examples* supporting the inquiries.

- AT/I must adequately *answer* the questions asked by the advisor and the user concerning existing definitions, strategy actions, hypotheses, values of statistical variables, etc.

- The advisor and the user must be able to *override the strategy* imparted so far, on single occasions, with an *order* for various purposes, such as to debug the strategy.

- The system should be able to assess the current domain situation in testing the effectiveness of a strategy action.

- A programmer should be able to *extend* the expertise of AT/I by including new domain-specific primitives without too much effort.

The operation of the AT/I system can be divided into two distinct parts:

- The *Learning* or *Knowledge Acquisition Phase* during which the system accepts advice from human experts interactively in the form of declarative and procedural knowledge, is taught some domain-dependent vocabulary, and alerts the advisor and the user of any effects modifications on one part of the strategy has on other parts;

- The *Operational Phase* during which the system applies the expert system constructed to a domain of interest, monitors its performance and suggests improvements to the strategy on the basis of its performance and the analysis of expert criticism.

4.2. The Approach

4.2.1. *The Learning Phase*

We present a few block diagrams in describing our approach. Figure 4.2.1 shows the principal program constituents active during the learning phase.

The *Advisor* interacts with AT/I through an *Advisor Interface* which sends Advisor input to the *Knowledge Manager and Verifier*. The latter may have to

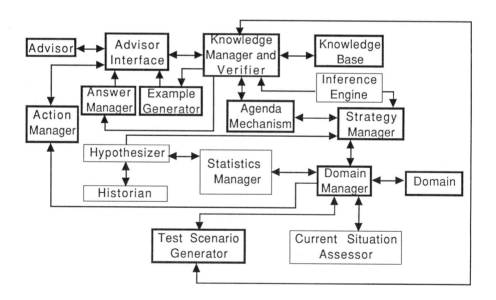

Fig. 4.2.1. — Block diagram of the principal components of the AT/I system — those active during the learning phase are marked with heavier lines

formulate clarifying questions aimed at the Advisor, again through the Advisor Interface. The Knowledge Manager and Verifier can also ask the *Example Generator* for help in trying to eliminate some ambiguity in an advice. The *Knowledge Base* can be accessed and modified only through the Knowledge Manager and Verifier. The *Agenda Mechanism* maintains a list of items that requires the advisor's attention at a later time. This happens when the advisor refers in his advice to an item prior to its definition or when he modifies a part of the strategy and this change causes some inconsistency in another part of the strategy. The *Answer Manager* is invoked when the advisor asks for a definition specified earlier or for an action prescribed by the strategy in a given situation (to which response the *Test Scenario Generator* may also contribute). The *Strategy Manager* receives information about and gives instructions to the *Domain* via the *Domain Manager*. The *Action Manager* communicates with the Advisor Interface and the Domain Manager about actions to be made and their effect.

A more detailed view of the Advisor Interface with its components is shown in Fig. 4.2.2. (Note that procedural knowledge is stored in some additional data structures that are of a more conventional type.)

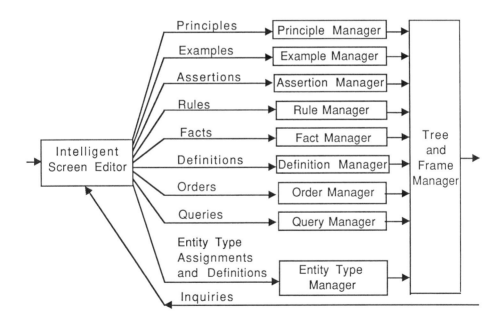

Fig. 4.2.2. — A detailed view of the Advisor Interface and its components

The advisor's input is transmitted via an *Intelligent Screen Editor*, a general purpose knowledge acquisition utility that is useful in accepting chunks of related input information. It offers an extensive 'help' facility to the advisor and the user; evaluates his answers for correctness; provides high-level editing capabilities for creating, retrieving and modifying individual data and whole information structures; enables the advisor and the user to postpone answers to certain questions or to select a default answer [185] for the lack of a better one. A given response to a question can indicate which question to ask next and what message should be generated for the advisor and the user. We note that changes within an editing session do not involve the Agenda Mechanism because they are contained locally in the editor. After the advisor leaves the editor and the information is transferred to the AT/I environment (to the Tree and Frame Manager), the Agenda Mechanism is activated if necessary.

Concerning the language of communication between the advisor or the user and AT/I, the following requirements can be stated:

- it should be easy to learn and use (no special programming skill should be needed to apply it to an established domain);
- production rules should be expressible in high-level, English-like statements;
- the system should be domain-independent yet able to acquire domain-specific vocabulary;
- the human expert should be able to modify what has been taught previously;
- a programmer should be able to extend AT/I code easily to include new domain-specific primitives.

The information stored can be divided into three categories:

- Some completely domain-independent basic modules and primitives;
- Structurally domain-independent trees and frames to be filled in by the advisor with domain-specific information;
- Domain-dependent vocabulary, variable names and ranges of values, default values, relations, basic processes, measures of performance, and the like.

The strategy is taught by the human expert in terms of both declarative and procedural advice. A measured level of redundancy is accepted in the knowledge representation to enhance processing. For example, the fact that two entities are related in some manner is stored with both entities. Further, production rules (expressing principles, examples and rules) are stored in both *interpreted* and *compiled* form. The interpreted version lends itself to internal analysis and manipulation but is not suited for execution. In contrast, the compiled form is fashioned for execution but cannot be easily examined by the AT/I itself.

4.2.2. *The Operational Phase*

The components active during the operational phase appear in Fig. 4.2.3 and will be discussed next. The AT/I continues to guide the development of an expert system when it is applied to a domain of interest. The *Inference Engine* accesses the declarative and procedural knowledge and meta-knowledge[1] acquired during the Learning Phase and employs it to produce a step toward the solution of the current problem. The *Current Situation Assessor* is

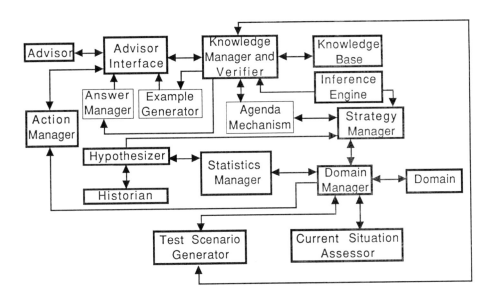

Fig. 4.2.3. — Block diagram of the principal components of the AT/I system — those active during the operational phase are marked with heavier lines

---

[1]As stated before, meta-knowledge refers to the information that characterizes the extent, quality, source, and cost of available knowledge.

responsible for evaluating the functioning of the expert system. The *Hypothesizer* suggests plausible new production rules or modifications to the strategy which will make the system more intelligent or efficient in areas deemed in need of improvement. Learning results in better performance through generalizations of existing advice, "tighter" compilation of domain-specific knowledge, and pruning of redundant, out-dated or unnecessary knowledge. The *Historian* module records problems and their solutions in support of the Hypothesizer. It can search performance records to discover previous problems which are similar to a given problem. The *Statistics Manager* also serves the Hypothesizer by doing the necessary data collection and evaluation.

The final debugging of the resulting expert system can usually be completed only in the operational phase. The *Test Scenario Generator*, the *Action Manager* and the *Current Situation Assessor* provide the task environment, the expert system response to it, and the evaluation of the consequences, respectively.

4.3.   The Implementation

4.3.1. *The Definition Facility*

The advisor is supposed to break up a complex domain into a collection of *contexts*. Simpler contexts are taught first and more complicated ones build upon what knowledge has already been imparted. The variables that appear in a context being dealt with (learned or applied) by the system are considered *active*.

Object definition [37] refers to the task of teaching the Advice Taker/Inquirer on the general structure of a class of objects — analogously to declaring a 'type' in Pascal. The advisor constructs a template that describes an object's make-up. (We adopt the convention to abbreviate the phrase "object's make-up" with the word "object" in the following.) Continuing the analogy, just as a programmer declares variables in terms of the defined types in Pascal, the advisor defines entities which are instantiations of the defined objects. The *Instantiation Facility* is described in Section 4.3.3.

The information structure chosen for representing objects is the *frame* [155]. The frame for a given object is a collection of information about the attributes that describe the object. Minimally, for each attribute, the frame contains knowledge about its data type, its range of permissible values and prototypical or default values which can be used if reasoning with incomplete knowledge is necessary.

The expert creates or modifies an object during the learning phase by entering the command **DEFINE** *<object>*. If a new object is being defined, the Intelligent Screen Editor displays an empty shell. As the advisor supplies information, additional questions relevant to the definition of the specific object are added to the screen. If an existing object definition is being modified, the system presents header information about the object and a list of its attributes. The advisor, by selecting an attribute, then causes the Intelligent Screen Editor to make its definition visible and available for being changed.

An object's frame consists of one or more attribute definitions. When only one attribute is necessary to convey all the relevant information about an object, it is more convenient to let the object's name refer to the value stored under the single attribute of that object. (For example, we could define the object *grocery bag* and the only attribute we may be interested in is its contents. Unambiguously, we can use the phrase "grocery bag" and understand its connotation to be "the contents of the grocery bag".) The name of the object serves as a descriptor for some quality about the object.

When only one attribute participates in the frame and the advisor decides that the object name should reference the value stored for that attribute, the class of object definition is termed a *descriptor;* otherwise, the class is known as a *record*. We also call an attribute-object pair a descriptor.

The advisor selects either a descriptor or a record for the object definition's class. If a record is chosen, the Intelligent Screen Editor requests the name of each attribute and its definition until the advisor has completed defining the object. If a descriptor is selected, the advisor is allowed to proceed directly with the definition.

The AT/I aids in constructing a type definition by furnishing broad type templates which the expert then customizes with respect to the attribute or descriptor being defined. The AT/I accepts numerical, ordered categorical, unordered categorical, list and advisor-defined types, which are next described.

The *numerical type* should be used if the value for the named attribute or descriptor is a single number or measurement. The advisor must in this case supply the lower and upper bounds of its range of values. The symbols -inf and +inf can be used for these if the range is open on either end or if there is no meaningful cut-off point. One can thus define a finite range, a range bounded at one end or a totally open range. Any constraints within the range (subintervals or points which are not to be included in the range of this definition) can also be specified.

In addition to supplying the range of values, the advisor is asked to provide, if such exists, the minimum measurable difference between two points along the range. This quantity, the Maximum Meaningful Resolution (MMR), characterizes the finite degree of precision of measuring the variable in question. (An advisor-supplied value, involving a higher degree of precision than defined, would indicate that the advisor has either misread the measuring instrument, mistyped the value, or committed some other error.)

Unless the quantity represented by this type is dimensionless, the complete type definition requires the unit of measurement used for the range and the MMR. (Information on converting from one unit of measurement to another can be supplied by invoking the Unit Definition Facility, described in Section 4.3.6.)

Finally, a numerical type definition requires English *expressions of comparison* which are synonymous with the phrases "is less than", "is equal to" and "is greater than". These context-specific phrases allow the expert to express his advice more naturally, aid in the parsing of production rules, and can be used by the expert system itself when explaining how or why a certain action has occurred.

The *ordered categorical type* should be declared if its range of possible values consists of a finite number of literal or symbolic values among which a meaningful ordering is possible (such as ranks in the military). This type is defined by furnishing all the possible values in the correct order, and supplying expressions of comparison for "is less than," "is equal to," and "is greater than."

The *unordered categorical type* is declared if the possible values are symbolic and can have no meaningful ordering among themselves (such as hair color). As in the previous case, the advisor must supply a list of all possible values. Expressions of comparison are requested by the system with the phrases "is equal to" and "is unequal to."

The definition is of the *list type* if the value for a named attribute or descriptor is a set. It is created by stating size of the set (if known) and the type of its elements (which need not be defined prior to referring to the elements).

In addition to customizing any type templates, the advisor can also define a type by using or modifying a type created previously, which is called the *advisor-defined type*. To do it, the name of the new type is declared and the name of an existing type is given. Once a frame has been defined, it becomes a valid type and can be used to define another object, an attribute which modifies an object, or a descriptor. Similarly, an existing definition of an attribute for some object can be referenced by entering **<*attribute*>** **OF** **<*frame*>** when asked for the existing type's name. Finally, the element type of a list type can be selected at any time by entering either **MEMBER OF** **<*descriptor*>** or **ELEMENT OF** **<*descriptor*>**, where *<descriptor>* is either a descriptor or an attribute of some object and is of the list type.

The expert can opt to *copy* or to *share* the definition of the chosen type. If the definition is copied, all references within the existing definition to itself are altered to refer to the new type in this version of the definition. If the definition is shared, a master-slave relationship is created between the existing and new type, and any changes made to the master type are propagated to its slaves. Synonymous types can also be created this way. The slave, however, cannot be altered independently of the master while the master-slave link is active. Any existing type, including a slave, can serve as the master to a new type. When a master type is altered, the slaves at all levels below it are affected. A slave can be freed from its bonds by changing the "share" facet of its definition to "copy". In this case, the definition of the absolute master is copied and references to its slaves are omitted.

Thus, an attribute can be defined in terms of another frame within a frame definition. This embedding of frames can occur up to any level — limited only by its utility and comprehensibility.

The expert can *modify an existing definition* by invoking the Intelligent Screen Editor for the object in question. If the object is a descriptor of not the advisor-defined type, the various properties known about the descriptor are displayed and made available for immediate modification. If the class of the object is a record, a list of its attribute names is displayed. By selecting an attribute, its definition is expanded on the screen and can be changed. If the definition of a descriptor or selected attribute is an advisor-defined frame, the Intelligent Screen Editor suspends editing at the current level of the object and presents the next lower level definition of the selected attribute or descriptor for modification. When editing at this lower level is exited, processing at the next higher level can be resumed.

A new attribute can be added to a record as well as an existing attribute can be deleted during an editing session, provided it is not the master in some master-slave relationship. Currently, the only (somewhat clumsy) way available to delete a master type definition is to copy its definition to one of its slaves and then transfer its remaining slaves to subservience under the copied definition. Deletion can then take place because the type no longer has any slaves dependent on its definition.

The frame structure, represented as a list of descriptors, is stored on the object's property list. In turn, information on the definition of each descriptor is stored on that descriptor's property list. Since the same attribute can have different definitions for different objects, a unique descriptor name is formed by concatenating the object name, a colon and the attribute name.

4.3.2. Prototypes

The *prototype for an object* is an ideal instantiation of the object, a composite view of all entities that are possible instantiations of the object. A prototype is a structure evoked to fill the gap when specific details are missing. Prototypes allow us to reason with default values when there is incomplete information for solving a problem.

The AT/I supports the creation and use of prototypes. To create a prototype, the command **PROTOTYPE** *<prototype>* is entered. The expert is then asked the name of the object that defines its structure. There is no limit on the number of prototypes that can be created from a given object. In effect, a prototype can be viewed as a stereotypical view for some subset of the entities that can stem from an object.

In defining the prototype, the advisor supplies default values for the attributes of the object. If it seems unnatural to specify an attribute with a value, none need be given. The process of interacting with the AT/I to fill in an object definition is described in detail in the next section.

4.3.3. *Entity Instantiation*

An entity in the AT/I is created by providing real values for the slots of a frame. The advisor invokes the Entity Instantiation Facility by issuing the command **ENTITY** *<entity name>*. If the entity already exists, its definition is made available for modification; otherwise, the system inquires which object definition this entity is to be patterned after and places the advisor in the Intelligent Screen Editor. The screen displays the name of the entity, the name of the frame, and writes the name of each attribute for which a value must be supplied on consecutive lines. The advisor can supply a value for a given slot, if it is known, or the slot can be left empty, which fact signifies that the value is not known or is to be computed by the system during the operational phase. The acceptable values for a given slot depend on the type of the template used to define the slot, which can be any of the types listed in Section 4.3.1. The following additional comments are in order:

- If the slot requires a measurement (the *numerical type*), the default unit of measurement is displayed and used if the advisor does not override it. However, the advisor can enter a numerical value and another unit of measurement, in which case the supplied measure is converted internally to the expected unit of measurement, provided the two units are of the same dimension (e.g., foot and meter). The converted value is checked to ensure that it falls within the acceptable range of values, and is not in a subinterval which has been excluded from this range. Finally, in case the MMR has been defined for this slot, a warning is generated if the advisor-supplied value contains more digits of precision than signified possible by the MMR.
- In case of *ordered and unordered categorical types*, the advisor must supply a value chosen from the domain of possible values for this slot.
- With the *list type*, the advisor must supply a list of values of the specified element type. Each value in the list is checked for validity. If the list is of fixed size, the number of values must also be given. It is permissible to repeat values within the list and the ordering used by the advisor is retained. If no value is entered for this list and a null list is legal in the context at hand, the advisor is asked whether this fact denotes that the list is empty or that the value is unknown.
- Finally, with the *advisor-defined type*, the name of an entity — which is the instantiation of the slot's type — must be entered. If the entity itself has not been created yet, the system generates a warning and internally

places the name of the supplied entity on a special structure called the *agenda* (see in Section 4.3.7). In general, an item is placed on the agenda if it is referred to by the advisor prior to its being defined. Any information the system can deduce about the undefined item is stored as notes. These notes are cross-referenced for consistency whenever an item is referred to elsewhere prior to its definition and at the time of its definition. If a conflict is detected, the advisor must select the correct usage of the item. All instantiations which incorrectly reference the item are marked erroneous and must be corrected. When a slot's value is flagged as invalid, the system functions as if the value were unknown.

4.3.4. *Production Rules*

The production rule is the basic form of procedural knowledge in the AT/I. Structurally, a production rule consists of an *antecedent* and a *consequent*. Only if the antecedent is true can the production rule be fired and, in turn, the consequent executed.

Specifically, a production rule has the form[1]:

IF *<clause>* [**AND** *<clause>*]* **THEN** *<statement>* [**AND** *<statement>*]*

Here *<clause>* is a logical test involving two expressions, such as "work process 15 takes more than 10 time units" or "assembly unit 8 has passed work station 11", and *<statement>* is a directive built out of AT/I primitive actions and domain-specific expressions previously defined for the AT/I. Figures 4.3.1, 4.3.2 and 4.3.3 provide a systematic explanation and examples.

When the advisor has defined a production rule, each clause and statement is verified both syntactically and semantically as soon as it is furnished. If no error is found, the system allows the advisor to continue with the definition of the production rule. However, if a clause or action does not parse, a detailed error message is generated and displayed in a pop-up window, and the erroneous part of the production rule is highlighted. The advisor can either correct the error right away or ignore it temporarily and furnish another part of the production rule. In any event, the entire production rule must parse prior to the system accepting it.

---

[1]An asterisk (*) as a superscript after a closing bracket (]) indicates that the contents of the preceding brackets may occur zero or more times whereas the plus sign (+) in the same position indicates an occurrence of one or more times.

4. THE ADVICE TAKER/INQUIRER

Substructures	Examples
literal	continuous
number	5
measurement	1.52 meter
descriptor	network
object	task
entity	chassis
function	center of gravity
function-arguments	center of gravity of the chassis
attribute-object pair	duration of the task
attribute-entity pair	weight of the chassis
attribute-function-arguments	successors of the first workstation

Fig. 4.3.1. — Substructures for creating expressions. Those of italicized names are also *decision variables* because they characterize the status of the environment and their values are the basis for deciding on subsequent actions

The syntactic parse is first performed and only if it succeeds can the semantic verification be undertaken. A definite clause grammar [174] is used to interpret the syntactic structure of the clauses and statements. The non-terminal nodes in the grammar are built-in and standard in the AT/I since the syntactical structure of the statements is fixed. In addition, new non-terminal nodes are necessary whenever the expert defines a domain-specific function or procedure because the grammar must incorporate the arity (the number of parameters) of the functions and procedures. The grammar automatically grows as the AT/I acquires new knowledge.

Terminal nodes are drawn from the set of advisor-defined object, entity, procedure and function names and their definitions, augmented by a small list of built-in functions and phrases of comparison.

The grammar is written so that any input will generate some parse. Further syntactic analysis of the parse either confirms the validity of the structure of the input or produces a list of errors therein. If errors are found, customized error messages are generated and displayed to the expert, and the invalid clause or statement is highlighted and internally flagged as requiring correction.

- *<Grammar for primitive action>*
- ° *<Example of usage of primitive>*

- *<variable>* = **CREATE ENTITY** *<object, entity,* or *prototype>*
- ° NEW CHART = **CREATE ENTITY** ASSEMBLY SEQUENCE

- *<variable>* = {**THE**} [**LAST** | **FIRST**] *<element>* [**OF** | **IN**] *<list>*
- ° FINAL ASSEMBLY OBJECT = **THE LAST** OBJECT **IN** ASSEMBLY SEQUENCE

- *<variable>* = *<elements>* **OF** *<list>* **THAT SATISFY** *<conditions>*
- ° HANDLED EVENTS = EVENTS **OF** OPERATIONS **THAT SATISFY**
  (START TIME OF EVENT IS NOT AFTER START TIME OF CURRENT TASK)

- *<variable>* = **PARTITION** *<list>* **ACCORDING TO** *<constraints>*
- ° QUALITY DISTRIBUTION = **PARTITION** MACHINED ITEMS **ACCORDING TO** QUALITY RATING

- **FOR EACH** *<element>* [**OF** | **IN**] *<list>* **PERFORM** *<actions>*
- ° **FOR EACH** CAR **IN** HALL #3 **PERFORM** (QUALITY CHECK, TEST RUN)

- **IF** *<conditions>* **THEN** *<actions>* **ELSE** *<actions>*
- ° **IF** (ENGINE TEMPERATURE IS HIGHER THAN 350 °C) **THEN**
  (CHANGE VALVE SETTING) **ELSE** (RUN ENGINE 50 HOURS MORE)

- **SORT** *<list>* **ON** [*<attribute>* [**INCREASING** | **DECREASING** | **+** | **-**]]$^+$
- ° **SORT** COMPLETED CARS **ON** COMPLETION TIME **INCREASING**

- **PUT** *<element>* **IN** *<list>*
- ° **PUT** MODEL X **IN** NO PRICE CHANGE

- **REMOVE** {**THE**} [**LAST** | **FIRST**] *<element>* **FROM** *<list>*
- ° **REMOVE THE LAST** WORK STATION **FROM** HALL #7

- **REMOVE** *<element>* **FROM** *<list>*
- ° **REMOVE** MODEL Y **FROM** NEXT YEAR'S PRODUCTS

- **EMPTY** *<list>*
- ° **EMPTY** ASSEMBLY SEQUENCE

- *<procedure name>* *<arguments>*
- ° COMPUTE WEIGHT OF MODEL Z

- **OUTPUT** [*<text string>* | *<evaluatable expression>*]$^+$
- ° **OUTPUT** "Weight of " Model Z " = " WEIGHT OF MODEL Z "."

Fig. 4.3.2. — Built-in AT/I primitives for constructing actions and some examples

```
+,  -,  *,  /
<,  <=,  =,  <>,  >=,  >
not
union
intersection
first
last
member
empty
partition
put
remove
sort
satisfy
size
```

Fig. 4.3.3. — Built-in AT/I functions for constructing production rules (the user can reference the relational operators by their English name)

Only after the syntactic parse is found error free does the system perform a semantic check on the input. The parse tree is examined to ensure that arguments to functions and procedures are of the correct type, phrases of comparison are used in the proper context, and units of measurement are consistent. If the semantic parse is also valid, the clause or statement is accepted; otherwise, suitable error messages are generated, and the input is highlighted and flagged as before.

The *compilation of a production rule* takes place when the expert has finished defining it and all input has been validated. The system encodes the production rule by transforming it from pseudo-English to Common LISP. Though this mapping is relatively quick, it is by no means trivial. References to objects are translated to LISP function variables, working variables are created for all local variables in the production rule, special encoding takes place to handle recursive function calls, and decision variables are represented in a canonical form so that the Inference Engine can evaluate or bind them as necessary.

A *decision variable* in the production rule can be defined with respect to an object or an entity. Any decision variable in the production rule, which is in terms of an object (i.e., there is no explicit mention of an entity), requires binding prior to the invocation of the production rule at run time. When the production rule is compiled into LISP, the object-based decision variables are translated into LISP code variables and placed in the parameter list of the resulting LISP function. This mapping is stored on the property list of the production rule, which is consulted by the Inference Engine prior to invoking the production rule during the execution phase.

A list of all the decision variables found in the antecedent is also stored as a property of the production rule. The *read-set* of the production rule is a combination of the decision variables in the antecedent and the independent decision variables found in the consequent of the production rule.

A list of the decision variables whose values are potentially altered if the production rule is fired — denoted the *write-set* of the production rule — is also placed on the property list of the production rule. In a similar fashion, the property list of each referenced decision variable is updated to indicate that this production rule can possibly alter the value of the decision variable. Every time a given decision variable occurs in a write-set, that production rule is added to a list saved on the property list of the decision variable. This list is denoted the *production rule packet* of the decision variable. The Inference Engine (see Section 4.3.8) accesses and examines the production rule packet for a given decision variable whenever the subgoal of instantiating or modifying that decision variable arises. Another list maintained for each decision variable is a collection of those production rules in which the decision variable occurs in the antecedent. This set of production rules is denoted the *production rule activation set*. Any production rule in this set contains an expression in its antecedent that is dependent on the given decision variable. Thus, if the value of the decision variable changes, it can possibly cause the antecedents of some production rules within its production rule activation set to be satisfied.

Knowing the read-set and the write-set of each production rule and which production rules are activated given a change in the environment, allows the implementation of an Inference Engine that has the capability of performing *forward and backward reasoning*. This additional meta-knowledge on the complexity of computing a value for a decision variable is used for creating an efficient reasoning mechanism. In summary, in order to decide on forward or backward reasoning, the Learning Phase of the AT/I analyzes procedural knowledge to discover

- which decision variables are instantiated as a result of executing a production rule (the write-set is the decision variables whose values are potentially altered when the production rule is fired);
- which decision variables are required for a given production rule to fire (the read-set is the decision variables in the antecedent and the independent decision variables in the consequent);
- which production rules can potentially instantiate a given decision variable (the production rule packet).

Modification of an existing production rule is performed by giving the command **RULE <*production rule name*>**. The production rule is displayed in the Intelligent Screen Editor and any portion of its definition can be altered, deleted, or expanded. The same syntactic and semantic verification occurs as described before. Once the production rule is accepted, its new encoding replaces the existing definition. The read-sets and write-sets are recomputed, and the production rule is removed from the property list of any decision variable which has appeared in the prior version of the production rule but no longer has a role in the modified version.

The production rule can also be deleted from the production rule set, in which case all references to it are removed from the knowledge base.

4.3.5. *Procedures and Functions*

The set of actions recognizable by the AT/I is expanded whenever the advisor creates new *procedures* or *functions*. The distinction between these two is the usual — a procedure call does not return a value and so it can be used only as a complete statement in the consequent of a production rule or within the body of a procedure or function; on the other hand, a function call returns a value and must be used as part of an expression in an antecedent or consequent.

The advisor can *define procedures* through the command **PROCEDURE <*procedure name*>**. The first step is to supply the parameter names and types of the procedure. Each parameter must be specified, and the order in which the parameters are given defines the order in which the parameters must be specified in any statement invoking the procedure. Procedures require a fixed number of parameters but it is also valid to define a procedure whose *arity* (the number of its parameters) is zero. The parameter type must be expressed in terms of an existing type of decision variables — object, descriptor, attribute-object pair, entity or attribute-entity pair. The ability to define the type of a parameter with respect to an entity is not conventional in computing practice but it is a convenience offered to the advisor and lends itself to a natural interpretation.

Procedure calls follow the call-by-name convention, so any assignment made to a decision variable, defined in terms of a parameter, changes the decision variable it is bound to at the next higher processing level.

The body of a procedure is defined the same way as the consequent of a production rule. The decision variables must be defined in terms of known entities or the procedure parameters. The procedure parameters are treated as the only valid objects recognizable within the procedure body. Recursion is fully supported, so a procedure can call itself either directly or indirectly.

The *encoding of a procedure* consists of adding its name to a list of advisor-defined procedures and functions, recording its arity and attaching a tag indicating that it is a procedure. The parameters and their required types are stored on the property list of the procedure and mapped into LISP variables for the actual LISP function definition. (The value returned by the LISP function is simply ignored by the AT/I since the encoding represents an advisor-defined procedure). The body of a procedure is encoded the same way in which a consequent of a production rule is encoded.

The write-set of the procedure is determined and placed on the property list. The list of entities and objects which are potentially activated through the execution of the procedure must be known so that the write-set of the higher level process invoking this procedure can be identified correctly.

If the definition of an advisor-defined procedure is modified, the old write-set of the procedure is replaced by the new write-set. If the arity of the procedure is changed or if the type of a parameter is modified, all production rules, procedures or functions containing calls to this procedure must be updated to reflect the change. The agenda mechanism tracks the required modifications. The AT/I flags any affected procedural knowledge and alerts the advisor of the required corrections. All flagged definitions are made inactive until the advisor remedies the incorrect procedure call and then the item is removed from the agenda.

The advisor can also delete an entire procedure, in which case the procedure name is removed from the list of advisor-defined procedures and functions. Also, the agenda mechanism records the production rules, procedures and functions whose definitions reference the purged function. Alternately, the advisor can supply a new procedure definition bearing the same procedure name. If the new parameter list has the same arity and order of parameter types, the flag set because of the deletion of the procedure is removed. The agenda mechanism oversees this task, too.

The advisor can *define and modify a function* through the command **FUNCTION** *<function name>*. As in a procedure definition, the advisor must specify the parameters of the function and their types but, in addition, the type of the value the function returns must also be given. Only entities, function parameters and the function name can be used as the basis for decision variables in the function body.

A problem arises when a call-by-name convention is adopted with respect to function calls. With a production rule control structure, the Inference Engine can test whether the left hand side of a production rule is satisfied prior to executing its right hand side. In doing so, a clause can be encountered which contains an expression containing an advisor-defined function call. Let us now assume that this function call involves certain parameters and that, as a result of executing the function, one of the parameters has its value changed. If the call-by-name convention is followed, the consequence is that the decision variable within the production rule bound to the function parameter is modified. However, if one of the clauses of the antecedent is subsequently discovered to be false, the Inference Engine would conclude that the production rule should not be fired and so it should have no effect on the environment at this time. This problem of erroneously changing the environment can be circumvented by knowing, prior to the evaluation of a production rule, the set of decision variables whose values might be modified as a direct result of executing the production rule. Fortunately, this information is readily available — it is stored as the write-set on the property list of the production rule. Thus, immediately before testing the antecedent of the production rule, the values of the decision variables in the write-set are saved. The antecedent is then tested and if it fails, the decision variables in the write-set are reset to their previous values, eliminating the unwanted side-effects described above.

The *encoding of a function* follows the same method as encoding a procedure. However, in addition, the return value of the LISP function is the value assigned to the advisor-defined function name. The list of advisor-defined functions and procedures is updated to include the function, the number of parameters and their types are saved on the property list of the function name, the write-set of the function is noted, and the definition is classified as an advisor-defined function. Finally, the routine described above for procedure modification is followed if any portion of an advisor-defined function is changed, including the type of the returned value.

4.3.6. *The Units of Measurement*

Reasoning about numerical information in complex domains is a common activity for expert system. Values of a given dimension can be provided in a large number of different *units of measurement*. The system must employ various scaling factors in converting a value within the same family of measurements (e.g., from foot to mile). Further, some dimensions, such as temperature, have conversion from one unit to another that is not only multiplicative but also involves the addition of a constant offset.

As discussed before, the advisor supplies the unit of measurement when defining a numerical type. The *Unit Definition Facility* enables the advisor and the user to define and use practically any unit for a value, including complex *derived units* which may involve a combination of metric families (e.g., newton = kilogram*meter/second^2). The command **UNIT** invokes the Unit Definition Facility. The advisor or the user can define a unit in the same family as another unit by entering a statement which gives the relative scaling factor between the two. Or else, a derived unit of measurement can be defined by entering the equivalent formula for which the unit is an abbreviation. The AT/I also requests the singular and plural forms of all units referenced in the Unit Definition Facility as shown in Fig. 4.3.4.

```
(a)  unit definition>   3 feet = 1 yard
        Is it true that 1 YARD equals 3 FEET?  yes
        What is the singular form of FEET?   foot
        What is the plural form of YARD?   yards

(b)  unit definition>  newton  =  kilogram*meter/(second*second)
        Is this conversion formula correct:
          NEWTON = (METER*KILOGRAM)/(SECOND*SECOND)?   yes
        What is the plural form of NEWTON?   newtons
        Is SECONDS the plural form of SECOND in this case?   yes
        What is the plural form of METER?   meters
        What is the plural form of KILOGRAM?   kilograms
```

Fig. 4.3.4. — Definitions of (a) two units in the same metric family, and (b) of a derived unit of measurement

Next we discuss how unit definitions are *encoded*. The singular and plural forms of every unit of measurement are kept on a list in the knowledge base. A list of conversion factors for switching to other units of measurement in the same family is maintained under each unit. When a new unit is learned in a particular family, the conversion factors for the units in the family are computed and the table of conversions is augmented to incorporate the new unit. For example, if the relation between foot and yard is first taught and this is followed by a statement on the relation between yard and mile, the AT/I immediately computes the scaling factor that links foot to mile.

Depending on how units are defined to the system, it is possible for the AT/I to not realize that what it is representing as two (or more) distinct families of measurement are really instances of subfamilies of the same metric family. Only if the advisor defines the conversion from a unit in one subfamily to a unit in another subfamily can the AT/I consolidate the two previously distinct groupings into one.

To aid the advisor, if the command **FAMILIES** is entered either at the AT/I command level or inside the Unit Definition Facility, the system generates a list of what it considers unrelated families of measurement. If the advisor notices that the system has separated units that should belong to one family, a definition must be supplied to unite them.

If a unit is an abbreviation for some combination of units, the formula that expresses this definition is stored on the property list of the unit. If more than one formula is given for a unit of measurement, only the most recently supplied formula is retained.

Formulae are represented in a canonical form, based on the observation that a derived unit represented in terms of other units can always be expressed in the form of a fraction where the numerator and denominator are products of units (or the value 1). Figure 4.3.5 shows some examples.

Derived Unit	Canonical Form
joule	(newton*meter)/1
newton	(kilogram*meter)/(second*second)
hertz	1/second

Figure 4.3.5. — Representation of the definitions of derived units in canonical form

A nice property of the canonical form is that it is closed under multiplication and division of units, so any derived unit expressed in terms of other units will result in a formula also in canonical form. This property permits elegant coding for computing conversion factors from any unit of measurement to any other one, including between different families of measurement. (The AT/I can find, for example, that to convert a measure of speed from feet per second to miles per hour, the value of speed should be multiplied by 15/22.)

Finally, we discuss how the AT/I uses the units of measurement. There are two areas in which the knowledge of the units of measurement is of importance during knowledge acquisition. The first is in the Entity Instantiation Facility whenever a numerical measurement is required. The system displays the default unit of measurement defined for the type of decision variable being instantiated. If the advisor supplies a value without qualifying it with a unit of measurement, the system assumes it is in terms of the default unit of measurement. However, if the advisor enters a value and a unit, the system first checks that the advisor's unit is in the required family and if it is not, the value is converted into the default unit of measurement automatically and tested against the valid range for the decision variable. The advisor is informed if the unit is not recognized or not known to be in the same family as the default unit of measurement.

The units of measurement are also employed during the parsing of defined production rules, procedures and functions. If a clause in an antecedent involves units of measurement, the two expressions connected by some relational operator must both evaluate to values that are in the same family of measurement. Similarly, if a decision variable is assigned a value as the result of executing some statement that contains dimensioned variables, both the decision variable and the expression determining its assigned value must be in terms of measurements from the same family. The advice is not accepted if it does not pass this uniformity requirement.

4.3.7. *The Agenda Mechanism*

The *Agenda Mechanism* is a data structure that maintains a list of items requiring the advisor's attention before the knowledge transfer can be completed. An item is placed on the agenda when

- the item is referred to in some other definition or advice prior to the definition of the item itself;
- the advisor has modified part of the strategy, which fact causes the use of this part to be inconsistent with its use elsewhere.

An entry on the agenda has the form:

(*<item> <reference> <notes>*)

Here *<reference>* is an identifier indicating where the *<item>* was referenced prior to its definition and/or inconsistency; *<notes>* carry specific information in the form of a list of property-value pairs about how the *<reference>* views the *<item>*.

The Agenda Mechanism aids in the acquisition of new domain-specific entities and vocabulary in giving the advisor and the user more flexibility, power and support in man-machine communication. It also displays partially completed definition templates and periodically reminds the advisor and the user of the entries to be taken care of.

4.3.8. *The Inference Engine*

We have described in Section 4.3.4 the meaning of the read-set and the write-set for a production rule, and that each decision variable within the production rule is tagged with reference to its role in it. We have also discussed the concepts of the production rule packet and the production rule activation set, and how they are used.

Given a decision variable whose value is unknown, the Inference Engine assembles all production rules that have write-sets referencing some aspect of the decision variable. Rules in this set fall into three categories, ordered from the most specific to the least specific in regard to the decision variable:

- production rules that specifically handle the given decision variable;
- production rules covering the entity itself if the decision variable is an attribute-entity pair;
- production rules governing the parent frame of the entity.

The production rule packets for these three cases are read directly from the property list of the named entity or frame — available from the computation during the knowledge acquisition phase. The production rules are merged into one list, with the order among the three categories retained. This list is referred to as the *potential production rule set*, which does not change during the operational phase. The *conflict-set* for the decision variable is initialized to this potential production rule set when the goal is first established through the creation of a goal node. The latter describes the problem and places it on a list of *outstanding goals*. A goal is removed from this list upon its achievement.

The Inference Engine then tries to identify the value of the decision variable. First, a production rule has to be selected from the conflict-set. A variety of heuristics have been employed to perform this conflict resolution. We have chosen a simple strategy which selects the most specific production rule over more general ones. This production rule is removed from the conflict-set. Whenever the conflict-set is empty, it indicates that the AT/I is unable to derive a value for the goal. The system in that case attempts to circumvent the need for a value of this specific decision variable by attacking the higher-level goal anew. (This can occur as long as its conflict-set is not empty; if it is, the next higher-level goal is reexamined, and so on). Only if the conflict-set for the top-level problem is empty does the AT/I admit defeat and turn to the user for help to be given in an interactive manner.

Assuming that a production rule has been selected from the conflict-set, the system then examines its *required bindings list* to discover what arguments the production rule needs, if any. This list of argument names and types is composed during the encoding of the production rule in the knowledge acquisition phase.

Provided the condition part of the production rule is satisfied, the action part is then executed. Here again, invocation of the Inference Engine is necessary if a statement cannot be evaluated because one of its arguments has no value. The same method outlined above is followed. When a statement is executed, one or more decision variables have their values affected. Any time there is a change in the state of the environment, additional production rules can potentially match the situation. Thus, whenever a decision variable is assigned a value, the activation set for that decision variable is used to refresh the conflict-sets of any outstanding goals. For a given goal, the activation set is compared against the potential rule set for that goal. Any rules found in common between the two sets are added to the conflict-set for the goal. The size of the conflict-set for any goal fluctuates dynamically as the state of the environment changes.

4.3.9. *The Current Situation Assessor*

The *Current Situation Assessor* module is responsible for evaluating the performance of the resulting expert system. Some of the evaluations contain numerical values combined with symbolic measures, such as: *as expected*, *potentially hazardous, no emergency anticipated, uneconomical* and *atypical*. The module can also be instructed to check whether a solution is obtained in a "reasonable period of time". The solution is not obtained if any of the following occurs:

- The system recognizes it cannot reach a solution because there are no applicable production rules to fire. The possible causes can be
    - the advisor has left part of the strategy incomplete (having ignored the warning given by the AT/I during the learning phase);
    - one or more decision variables have been defined incorrectly as having too small a range of possible values;
    - some action has been defined incorrectly or incompletely;
    - the problem is not solvable within the framework specified.
- The certainty factor computed for the solution is too low (cf., [201]). The possible reasons can be
    - the problem contains too many unknown or uncertain decision variable values;
    - the strategy of the advisor is not effective enough in the domain at hand.
- The system is not converging to a solution. The possible reasons for it can be
    - infinite looping;
    - the system is unable to "focus its attention".

If a solution is recognized as being of poor quality by the Current Situation Assessor, the Hypothesizer (described in Section 4.3.10) is informed and a *case history* is drawn up for analysis. The case history contains a trace of the system's attempt to reach a solution, a table of the known variables in the environment and their values, and a set of similar problems solved successfully in the past (provided by the Historian, described in Section 4.3.11).

4.3.10. *The Hypothesizer*

The *Hypothesizer* module examines the trace of the attempted solution by the system and tries to isolate that part of the strategy which is at fault. The tentative reason for the system's failure, provided by the Current Situation Assessor, guides the Hypothesizer in its search. The possible causes can be as follows:

- An *infinite recursion* occurs when the system is unable to produce a change in variable values that would fulfill some terminating condition to halt looping. The Hypothesizer needs to isolate the minimal set of contexts involved in the infinite recursion. Then it must discover whether some set of decision variable values can achieve the terminating condition. If this is not possible, the strategy needs to be changed so that a terminating condition can be reached. If some set of values eventually succeeds in producing it, the Hypothesizer needs to find out why these values were not achieved in the first place. If certain specific values are required for the terminating condition to succeed, it is useful to see how close the system came to reaching these. It is also possible that the terminating condition is too specific.
- The *Certainty Factor* may be too low. If the majority of the active decision variables have known values but the Certainty Factor is low, it indicates bad logic in the suggested solution strategy. The Hypothesizer should try to find a production rule or set of production rules which cause the Certainty Factor to decrease. If too many variables had unknown values when the problem was presented to the system, one can assume nothing about the quality of the logic. The problem in this case is to discover why so many variable values were not known. The case is then not the fault of the expert system and the advisor is to be consulted by the AT/I about some revision of the advice.
- If a *dead end* occurs with the expert system when it is unable to find a solution after having exhausted its knowledge available, the production rules used may be too specific. The Hypothesizer should in that case suggest relaxing some of the production rules constraints. The Hypothesizer needs to check how close the decision variable values have come to satisfying the condition clause of the failed production rule and whether it is it possible to extend the respective range specified so that it can succeed.

4.3.11. *The Historian*

The *Historian* module records problems and their solutions for future reference. The information stored consists of a list of active variables and their values, and an ordered list of production rules fired to achieve the solution.

The Historian searches the archives to discover previous problems which are similar to a given problem. The following set of factors affect the degree of similarity:

- do the same variables have known values;
- if the same variable is numerical, do its respective values cover the same or similar percentage of the total range. (For example, if the range is [0,100] and there are two values 25 and 30 in one problem, and 35 and 40 in the other; these pairs of values are within 5% of the total range);
- if a variable is ordered categorical, are the values adjacent;
- if a variable is unordered categorical, are the values the same.

Based on such factors, a *metric of similarity* can be devised with a value between 0 and 1. If two problems have a similarity rating greater than some threshold value, the Historian considers them similar and provides this information to the Hypothesizer.

4.4. Benefits of Using the AT/I

The AT/I creates an expert system by interacting with an advisor who defines a strategy for discovering solutions in his domain of interest. The advisor and the user benefit in a variety of ways:

1. The advisor must formalize his knowledge, which will enhance his own understanding of the domain. The explicit codification of knowledge, usually implicit and elusive, leads to new insights and ideas, and may eliminate incorrect concepts.

2. The AT/I is able to detect and help correct areas in the advisor's strategy in which his knowledge is inconsistent, redundant and/or incomplete.

3. The successful creation of an expert system reduces the shortage of experts in the advisor's domain and frees human experts to pursue those activities which a computer cannot do (as yet).

4. The AT/I can accept knowledge from several advisors and, by selecting the best features from each of the advisors' strategies (as discussed in Chapter 3 on the Quasi-Optimizer), it can produce a super strategy which outperforms them all.

5. A student or trainee could teach his strategy to the AT/I, which could then be compared with the expert's strategy to discover differences or viable alternatives.

6. The AT/I could be used to present problems to a student and to judge the validity of the student's response, either by comparing it against the expert's strategy or by using the student's solution and following it through to discover its strengths or weaknesses.

7. A student could study the strategy produced by the AT/I and question its various aspects, which facility provides a novel, interactive method of learning.

A preliminary prototype AT/I system has been electronically connected to the Simulated Air Traffic Control (ATC) environment (see Chapter 8) to investigate the system requirements for the operational phase. There are several roles in that exemplary domain that the AT/I can assume (when the program is appropriately modified):

- The *controller* — one or several experienced ATC operators advise the AT/I about the appropriate decision making strategy and the resulting expert system can act, as a controller, in the real-life environment.
- The *assistant* — the AT/I is taught a subset of the ATC decision making strategy which is less critical in nature. The resulting expert system, as an assistant, handles the corresponding aspects of the task while letting the human operator be responsible for the more critical aspects. The division of labor may in fact be dynamically determined by the human operator, depending on his current load.
- The *trainer/critic* — an expert ATC operator advises the AT/I and the resulting expert system, as a trainer/critic, monitors and critiques the activity of ATC trainees.
- The *trainer/simulator* — the AT/I is advised by an ATC trainee and the resulting expert system, as a trainer/simulator, controls a simulated ATC environment for the educational benefit of the trainee.

4.5. Applications in Assembly Line Balancing and Street Traffic Light Control

This Section demonstrates the ability of the AT/I to acquire and use domain-specific knowledge. The first area of application is a major phase of assembly line balancing. We first define the problem area and then various scenarios in the knowledge acquisition phase are shown as the AT/I learns the expert's strategy. Finally, an exemplary run of the operational phase is given and the AT/I's solution is analyzed.

The second application under development concerns the teaching of strategies for a single street traffic light control mechanism. (A *distributed control regime* for such a task is discussed in Section 6.5.)

4.5.1. *The Definition of the Assembly Line Balancing Problem*

The assembly line is normally divided into distinct workstations, at each of which a certain set of operations is continually performed. An individual workstation may be responsible for one or more tasks in the manufacturing of a product. The assembly line balancing problem concerns the identification of an optimal (or near-optimal) configuration and sequencing of workstations, such that some product can be assembled given certain time and ordering constraints.

Stated more formally, in order to assemble a product, a known set of tasks must be performed. The operation time per unit product is known for each task and a known partial ordering relationship exists among the tasks. The problem is to compute an arrangement of workstations and their compositions (task assignments) such that

- each task is assigned to one and only one workstation;
- the sum of the times of all tasks assigned to any one station does not exceed some maximum (the *cycle time*);
- the stations formed can be ordered so that the partial orderings among tasks are not violated;
- the number of workstations formed is minimized;
- the total (or the maximum) idle time of workstations is minimized.

A heuristic method for approximating optimum solutions to the assembly line balancing problem is given in an early AI work by Tonge [214]. He divides the problem into three phases:

(1) constructing a hierarchy of increasingly simpler line balancing problems;
(2) grouping tasks into workstations; and
(3) "smoothing" the resulting balance.

We have selected Phase I, thought to be of suitable complexity to demonstrate various features and abilities of the AT/I.

Phase I collects elements (unit tasks) that either share the same predecessors and successors or that occur in an uninterrupted, unbranching sequence. Correspondingly, one process in Phase I has the responsibility to create *sets*, and another one to form *chains*, as shown in Fig. 4.5.1. Phase I consists of repeated formations of *complex elements* (sets and chains), each of which replaces a group of elements in the network with its equivalent complex element until the network cannot be reduced any further. Phase I is successful if the entire network can be transformed into a single complex element.

As a first step of teaching the strategies involved to AT/I, we must identify all objects in the domain and describe their structures. We then define the functions and procedures useful for the problem and specify a set of production rules. Finally, we present a run of the resulting mini-expert system.

The low-level element in assembly line balancing is the *task*. The only property of a task relevant now is *duration*. Figure 4.5.2 shows the creation of

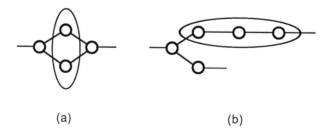

(a)   (b)

Fig. 4.5.1. — A *set* (a) and a *chain* (b) of elements

```
AT/I Command:  DEFINE  TASK

Frame name:    TASK
     Class:    DESCRIPTOR or RECORD

     Attribute:  DURATION
     Type:       NUMERICAL, ORDERED CATEGORICAL, UNORDERED
                 CATEGORICAL, LIST,  USER-DEFINED
                 — Acceptable range of values  —
                 Lower bound:  0
                 Upper bound:  +INF
                 — Constraints on range  —
                 Are there any constraints on this range? NO or YES
                 Maximum meaningful resolution:  1
                 — Unit of measurement  —
                 Unit of measurement (singular):  MINUTE
                 What is the plural form of MINUTE?  MINUTES
                 — Expressions of comparison  —
     <           IS  SHORTER  THAN
     =           IS  AS  LONG  AS,  IS  THE  SAME  AS,  IS  AS  SHORT  AS
     >           IS  LONGER  THAN

     Attribute:

OPTION:
```

Fig. 4.5.2. — The definition of the frame TASK. Here and in the diagrams below, words in **bold face** are inputted by the user, words in bold face and shadow are chosen by the user out of several options

the object TASK with the AT/I, using the Intelligent Screen Editor. A given collection of tasks is assigned to a *workstation*. Attributes of a workstation are *duration*, *type* (unit, set or chain), *successors* and *predecessors* (to represent the knowledge on partial ordering) and, of course, *tasks* (those associated with the workstation at hand).

Figure 4.5.3 shows a portion of the object definition for WORKSTATION. The group of some or all workstations necessary to produce a product is referred to as the *network*.

```
AT/I Command:  DEFINE  WORKSTATION

Frame name:    WORKSTATION
       Class:  DESCRIPTOR or RECORD

       Attribute:  TYPE

       Attribute:  DURATION

       Attribute:  SUCCESSORS

       Attribute:  PREDECESSORS

       Attribute:  TASKS
       Type:   NUMERICAL, ORDERED CATEGORICAL, UNORDERED
       CATEGORICAL, LIST, USER-DEFINED
                   — Definition of list —
       Singular form of TASKS:  TASK
       Type of element in list:   NUMERICAL, ORDERED CATEGORICAL,
                  UNORDERED CATEGORICAL, USER-DEFINED
       Name of user-defined type:  TASK
       Size of list (enter * if no fixed size):  *

       Attribute:

OPTION:
```

Fig. 4.5.3. — A portion of the definition of the frame WORKSTATION

The definition of the NETWORK, showing also two other descriptors, is given in Fig. 4.5.4. This diagram shows various uses of the **VIEW** command.

Given a network, our mini-expert system continues to group chains and sets of workstations together until it has either reduced the network to a single workstation, or no more sets or chains exist. Two advisor-defined functions are useful: *SET REDUCTION* and *CHAIN REDUCTION*. When applied to a network, either of these functions returns a literal indicating whether a set or chain exists, respectively. If reduction of the network is possible, the respective procedures *REDUCE SET* and *REDUCE CHAIN* can be invoked. *REDUCE SET*, given a network and a set of workstations within the network, modifies the network so that the set of workstations is assembled into a new workstation, and the predecessors and successors of the affected workstations in the network are adjusted. *REDUCE CHAIN*, given a network and

```
AT/I Command:  VIEW  NETWORK
DESCRIPTOR:    NETWORK
     Type:     LIST of unspecified size of WORKSTATION

AT/I Command:  VIEW  OUTCOME
DESCRIPTOR:    OUTCOME
     Type:     Unordered categorical
     Possible values:  POSSIBLE, IMPOSSIBLE.
     English relationships:
               =   − not supplied by expert −
               <> − not supplied by expert −

AT/I Command:  VIEW  COLLECTION
DESCRIPTOR:    COLLECTION
     Type:     LIST of unspecified size of NETWORK

AT/I Command:   VIEW FRAMES
     −          COLLECTION
     −          OUTCOME
     −          NETWORK
     −          WORKSTATION
     −          TASK
```

Fig. 4.5.4. — The **VIEW** command

the lead-in workstation of a chain, groups the workstations of the chain into a new workstation, and modifies the network to reflect the new configuration.

The function *GATHER CHAIN* is used by *REDUCE CHAIN* to collect the workstations that define the chain. The high-level definition of the function *SET REDUCTION* is shown in Fig. 4.5.5.

There are four clusters of *production rules* for this particular phase of assembly line balancing. First, a couple of rules check whether a solution has been found yet, is still possible or is impossible. These rules are shown in Fig. 4.5.6. Two production rules are used to compute the duration of a workstation as indicated by Fig. 4.5.7.

```
Function name:  SET REDUCTION
Function return type:  OUTCOME

== Parameters ==
     Parameter name:  NET
     Parameter type:  NETWORK

== Local variables ==
     Local variable name:  GROUPING
     Local variable type:  COLLECTION

== Statements ==
*)   GROUPING = PARTITION THE NETWORK ACCORDING TO
     (SUCCESSORS, PREDECESSORS)
*)   IF SIZE OF GROUPING IS LESS THAN THE SIZE OF THE NET
     THEN SET REDUCTION = POSSIBLE
     ELSE SET REDUCTION = IMPOSSIBLE
*)
Option:  EXIT
```

Fig. 4.5.5. — Schema for the high-level definition of a function

The remaining rules control the network reduction process. Only one rule is needed to initiate set reduction (see Fig. 4.5.8). Three production rules are used to guide chain reduction (see Fig. 4.5.9).

```
RULE-3
== IF ==
*)   THE SIZE OF THE NETWORK IS EQUAL TO 1
== THEN ==
*)   ROOT = THE NETWORK
*)   STATUS = POSSIBLE

RULE-4
== IF ==
*)   THE SIZE OF THE NETWORK IS GREATER THAN 1
*)   SET REDUCTION OF THE NETWORK IS IMPOSSIBLE
*)   CHAIN REDUCTION OF THE NETWORK IS NOT POSSIBLE
== THEN ==
*)   STATUS = IMPOSSIBLE
*)   ROOT = THE NETWORK
```

Fig. 4.5.6. — Production rules to decide if a solution has been reached

```
RULE-1
== IF ==
*)   THE  TYPE  OF  WORKSTATION  IS  EQUAL  TO  UNIT
== THEN ==
*)   FOR  EACH  TASK  IN  THE  TASKS  OF  THE  WORKSTATION   PERFORM
     SUM  =  SUM  +  DURATION  OF  THE  TASK
*)   DURATION  OF  WORKSTATION  =  SUM


RULE-2
== IF ==
*)   THE  TYPE  OF  WORK  STATION  IS  NOT  EQUAL  TO  UNIT
== THEN ==
*)   SUM  =  0  MINUTES
*)   FOR  EACH  STATION  IN  THE  STATIONS  OF  THE  WORK  STATION
     PERFORM
     SUM  =  SUM  +  DURATION  OF  THE  STATION
*)   DURATION  OF  WORK  STATION  =  SUM
```

Fig. 4.5.7. — Rules for computing the duration of a workstation

```
RULE-5
== IF ==
*)   SET  REDUCTION  OF  THE  NETWORK  IS  POSSIBLE
== THEN ==
*)   GROUPING  =  PARTITION  THE  NETWORK  ACCORDING  TO
     (PREDECESSORS,   SUCCESSORS)
*)   FOR  EACH  ELEMENT  IN  GROUPING  PERFORM
     IF  SIZE  OF  ELEMENT  IS  GREATER  THAN  1
     THEN  (OUTPUT  "Reduce  Set  of"  ELEMENT  "and"  NETWORK,
     REDUCE  THE  SET  OF  ELEMENT  AND  THE  NETWORK)
```

Fig. 4.5.8. — The rule for set reduction

```
RULE-6
== IF ==
*)   A SITE IN NETWORK HAS NO PREDECESSORS
*)   THE SITE HAS 1 SUCCESSOR
*)   THE SUCCESSOR OF THE SITE HAS ONLY 1 PREDECESSOR
== THEN ==
*)   OUTPUT "Reduce Chain of" SITE "and" NETWORK "."
*)   REDUCE CHAIN OF THE SITE AND THE NETWORK

RULE-7
== IF ==
*)   A SITE IN NETWORK HAS MORE THAN 1 PREDECESSOR
*)   THE SITE HAS ONLY 1 SUCCESSOR
*)   THE SUCCESSOR OF THE SITE HAS ONLY 1 PREDECESSOR
== THEN ==
*)   OUTPUT "Reduce Chain of" SITE "and" NETWORK "."
*)   REDUCE CHAIN OF THE SITE AND THE NETWORK

RULE-8
== IF ==
*)   A SITE IN THE NETWORK HAS 1 PREDECESSOR
*)   THE SITE HAS 1 SUCCESSOR
*)   THE PREDECESSOR OF THE SITE HAS MORE THAN 1 SUCCESSOR
*)   THE SUCCESSOR OF THE SITE HAS 1 PREDECESSOR
== THEN ==
*)   OUTPUT "Reduce Chain of" SITE "and" NETWORK "."
*)   REDUCE CHAIN OF THE SITE AND THE NETWORK
```

Fig. 4.5.9. — Rules for chain reduction

Let us now look at the *operational phase* from a user's point of view and assume that the network is as shown in Fig. 4.5.10. The system created during the first phase of assembly line balancing uses the entities ROOT (of type *network*) and STATUS (of type *outcome*) to return the results. The user must initialize these to have unknown values (see Fig. 4.5.11) and instantiate all tasks involved in the manufacturing of a given product. The user initially creates a workstation to include one of every task and specifies the partial ordering among these elements. Finally, the network itself must also be also instantiated (see Fig. 4.5.12). Once this preliminary set-up is completed, the system can be directed to compute the root of the network if it exists (see Fig. 4.5.13).

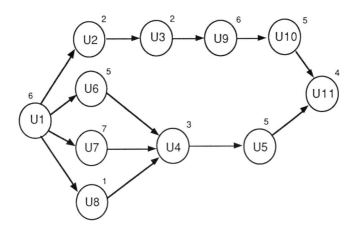

Fig. 4.5.10. — Initial configuration for an eleven element assembly line balancing problem; task names are shown within the circles and the small numbers next to the circles indicate the durations of individual task accomplishment; in the first iteration, each task will be assigned to a separate workstation

```
AT/I Command:   ENT  ROOT

ROOT is a specific example of which frame?   NETWORK

Entity name:    ROOT
Parent frame:   NETWORK

NETWORK:        UNKNOWN, EMPTY LIST

Option:         EXIT
```

Fig. 4.5.11. — The value of ROOT is set to unknown at first

```
AT/I Command:   ENT U1

U1 is a specific example of which frame?   TASK

Entity name:    U1
Parent frame:   TASK

TASK DURATION [MINUTES]:   6 MINUTES

Option:         EXIT


AT/I Command:   ENT W1

W1 is a specific example of which frame?   WORKSTATION

Entity name:    W1
Parent frame:   WORKSTATION

WORKSTATION
     TYPE:      TASK
     DURATION [MINUTES]:   6 MINUTES
     SUCCESSORS:   W6, W7, W8
          The entities {W6, W7, W8} have not been defined yet.
          Are these undefined entities of type WORKSTATION? YES, NO
     PREDECESSORS:     UNKNOWN, EMPTY LIST
     TASKS:     U1
     STATIONS:         UNKNOWN, EMPTY LIST
Option:         EXIT


AT/I Command:    ENT ELEVEN ELEMENT PROBLEM

ELEVEN ELEMENT PROBLEM is a specific example of which frame?
                NETWORK

Entity name:    ELEVEN ELEMENT PROBLEM
Parent frame:   NETWORK

NETWORK:        W1, W2, W3, W4, W5, W6, W7, W8, W9, W10, W11

Option:         EXIT
```

Fig. 4.5.12. — User instantiation of entities

```
Goal:  the value of ROOT

>>> Trying RULE-4 ...
     Please supply a value for NETWORK:  MY NETWORK
>>> Trying RULE-3 ...
>>> Trying RULE-8 ...
Reduce Chain of W2 and (W1 W2 W3 W4 W5 W6 W7 W8 W9 W10 W11)

     == goal:  the DURATION of W2 ==
     >>> Trying RULE-2 ...
     >>> Trying RULE-1 ...
     >>> Successful:  the DURATION of W2 is 2 MINUTES <<<

     == goal:  the DURATION of W3 ==
     >>> Trying RULE-2 ...
     >>> Trying RULE-1 ...
     >>> Successful:  the DURATION of W3 is 2 MINUTES <<<

      ...  { DURATION of W9 and DURATION of W10 found }

>>> Trying RULE-5 ...
Reduce Set of (W6 W7 W8) and (W1 W4 W5 W6 W7 W8 W11 ENT-387)

     ...  { DURATION of W6, DURATION of W7, DURATION of W8 found }

>>> Trying RULE-5 ...
>>> Trying RULE-6 ...
>>> Trying RULE-7 ...
>>> Trying RULE-8 ...
Reduce Chain of ENT-392 and (W1 W4 W5 W11 ENT-387 ENT-392)

     ...  { DURATION of W4 and DURATION of W5 found }

>>> Trying RULE-5 ...
Reduce Set of (ENT-387 ENT-396) and (W1 W11 ENT-387 ENT-396)

>>> Trying RULE-5 ...
>>> Trying RULE-6 ...
Reduce Chain of W1 and (W1 W11 ENT-399)

     ...  { DURATION of W1 and DURATION of W11 found }

>>> Trying RULE-5 ...
>>> Trying RULE-6 ...
>>> Trying RULE-7 ...
>>> Trying RULE-8 ...
>>> Trying RULE-3 ...
>>> Successful:  the value of ROOT is (ENT-400) <<<
```

Fig. 4.5.13. — The expert system computes the root of a network

The contents of the workstations created by the expert system can be viewed with the **VIEW** command (see Fig. 4.5.14). For this particular example, five workstations have been created.

```
AT/I Command:  VIEW ENT-400
Entity:        ENT-400 (of type WORKSTATION)
    TYPE:      CHAIN
    DURATION: 46 MINUTES
    SUCCESSORS:     { }
    PREDECESSORS:   { }
    TASKS:     U11, U1, U4, U7, U6, U8, U5, U9, U2, U3, U10
    STATIONS: W1, ENT-399, W11

AT/I Command:  VIEW ENT-399
Entity:        ENT-399 (of type WORK STATION)
    TYPE:      SET
    DURATION: 36 MINUTES
    SUCCESSORS:     W11
    PREDECESSORS:   W1
    TASKS:     U4, U7, U6, U8, U5, U9, U2, U3, U10
    STATIONS: ENT-387, ENT-396

AT/I Command:  VIEW ENT-387
Entity:        ENT-387 (of type WORK STATION)
    TYPE:      CHAIN
    DURATION: 15 MINUTES
    SUCCESSORS:     W11
    PREDECESSORS:   W1
    TASKS:     U10, U3, U2, U9
    STATIONS: W2, W3, W9, W10

AT/I Command:  VIEW ENT-396
Entity:        ENT-396 (of type WORK STATION)
    TYPE:      CHAIN
    DURATION: 21 MINUTES
    SUCCESSORS:     W11
    PREDECESSORS:   W1
    TASKS:     U5, U8, U6, U7, U4
    STATIONS: ENT-392, W4, W5

AT/I Command:  VIEW ENT-392
Entity:        ENT-392 (of type WORK STATION)
    TYPE:      SET
    DURATION: 13 MINUTES
    SUCCESSORS:     W4
    PREDECESSORS:   W1
    TASKS:     U8, U6, U7
    STATIONS: W6, W7, W8
```

Fig. 4.5.14. — Five workstations created by the expert system

4.5.2. *The Definition of the Street Traffic Light Control Problem*

The control of street traffic lights by computers is a problem of significant importance from the point of view of economy and safety. We are developing a simple AT/I-generated expert system that can control the traffic light at an intersection, in response to spatially and temporally local data on cars arriving at an intersection. (A more sophisticated system also under development, employing the paradigm of distributed planning and problem solving, is described in Section 6.5.)

Let us consider Fig. 4.5.15. Eight sensors (S) report to the processor (P) the number and the speed of cars arriving at the left-turn lanes and the continuing/right-turn lanes in the four directions. On the basis of these data, the processor controls the parameters of the traffic light cycle. The objective is to minimize the average (or the maximum) waiting time of the cars at the intersection, regardless of how the local decisions affect the traffic pattern at the adjacent and farther intersections.

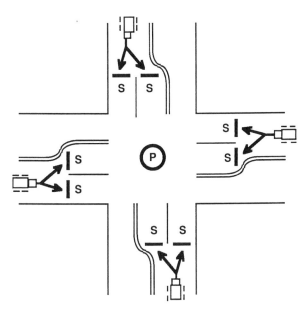

Fig. 4.5.15. — A schematic representation of intersection: each of the left-turn lanes and the continuing/right-turn lanes are equipped with a sensor (S) sending data about cars to the processor (P) which controls the traffic light

The advisor has to define the following[1]:

- a *left-turn lane* having the properties of *traffic direction* (N, E, S, W), *maximum capacity* of cars waiting in it and *current contents* (an indefinite number of cars currently waiting in it, which must be less than or equal to the maximum capacity);
- a *continuing/right-turn lane* having the properties of *traffic direction* (as above), *current contents* (the number of cars *known* to be waiting in it[2]), *speed limit* in the street direction, *current traffic flow* in it (number of cars going over the sensor per unit time), *jam-length* (time period during which no car has moved over the sensor);
- traffic light parameters (*total cycle length, starting time* of the cycle in one street direction, the *length of the green light* and the *length of the yellow light* in the two street directions, the *length of the left-green-arrow light* in all four traffic directions);
- *congestion* and *severe congestion* (in which case not every car known to be waiting in a direction can get through the intersection during the next cycle or during the next two cycles, respectively);
- *direction1* is any traffic direction (N, W, S or E), *direction2, direction3,* and *direction4*, each rotated clockwise by 90° from the previous one;
- traffic is headed in *one street* in direction1 or direction3, and in the *other street* in direction2 or direction4;
- *total traffic flow* in one street direction is the sum of the corresponding traffic flows through the two left-turn lanes and the two continuing/right-turn lanes;
- *average* or *maximum waiting time* of cars in a given direction — depending on which measure is to be minimized.

The production rules are to be specified in a parametrized form so that they can be optimized during the operational phase. A few exemplary production rules to be defined are shown next.

- **if** the *total traffic flow* in *one street* is greater than X (="busier street") **and** in the *other street* is less than Y (="less busy street") **then switch** to *semiactuated strategy*;

---

[1]There is a distinction between 'street direction' (going North-South or East-West) and 'traffic direction' (North, East, South or West).

[2]The maximum of this measure is determined by the distance between the sensor and the (imaginary) line of the intersecting street divided by the shortest length of a car. In other words, the processor does not know about cars that *may* be waiting before they pass over the sensor.

- **if** under *semiactuated strategy*
  and *current contents* of *left-turn lane* **plus** *current contents* of
      continuing/right-turn lane in *less busy street* is less than *Z*
  and *clock time* of *green light* in *busier street* is less than *T*
  then do not **switch** light in *busier street* to yellow;

- **if** there is no *congestion* or severe *congestion* in either street
  then set the ratio between the *length of the green light* in *one street* and
  the *length of the green light* in the *other street* **equal** to the ratio between
  the *total traffic flow* in *one street* and the *total traffic flow* in the *other
  street*;

- **if** *yellow light* in *direction 1* and *direction3* is about over
  and *current contents* of the *left-turn lane* in *direction2* is not **equal** to 0
  then switch light in *direction2* to *green-left-arrow*;

- **if** there is *congestion* in *one street*
  and there is no *congestion* in the *other street*
  then set the *length of green light* in *one street X* seconds longer;

A related graphics program called "Roads Scholar" has been developed,
which will enable the advisor and the user to evaluate, also in the visual
mode, the effect of changes in the parameters, composition and hierarchy of
the rules used.

4.6. Summary

The Advice Taker/Inquirer is a domain-independent system that is capable of
learning and applying an expert's strategy. The system is logically divided
into two phases: the knowledge acquisition phase, during which a human
advisor interactively defines his strategy for solving problems; and the
operational phase, during which the expert system constructed is applied to a
domain of interest.

Declarative knowledge is transferred to the AT/I in terms of object
definitions and entity instantiations. An object definition is a general
description of the characteristics that define some class of objects. The
advisor also supplies the AT/I with a domain-specific vocabulary covering
the essential attributes of the object, phrases for comparing objects in the
same class and units of measurement for the various qualities of the object.

In addition, default values typical of the object can be supplied for different contexts (prototypical knowledge). Entities within the environment are created by supplying values to an object template.

Procedural knowledge is expressed in terms of production rules, functions and procedures. Various enhancements to the standard production rule programming paradigm are used to capture the advisor's strategy in a more natural form. Local variables can reduce the number of production rules necessary to express a piece of advice. The consequent to a rule is specified by using statements in a high-level, English-like AT/I language. The advantages of this language include the ease of its usage, the powers of expressibility, extensibility, error handling, and compilability.

All advice is strongly type checked, and special attention is given to the advisor's ability to modify his strategy. The system automatically reports the effects that a change to a part of the strategy has on other parts and monitors the need for related modifications through an agenda mechanism.

The operational phase allows the user to pose questions to the resulting expert system and to contribute to its improvement. The built-in inference engine applies its meta-knowledge about the relations between variables and production rules to enable both forward and backward chaining in its reasoning process.

4.7.   Acknowledgements

Bob Cromp has done the lion's share of this work. We also acknowledge our debt to George Sicherman, Steve Feuerstein, Skip Lewis and the other members of our Group for Computer Studies of Strategies for their contributions of ideas, work and criticism.

5. The Generalized Production Rule System (GPRS)

5.1. Introduction and Research Objectives

In decision making, a fairly precise knowledge of the values of all relevant situational variables is necessary. We call *open variables* (OVs) those that can be observed and measured at any space and time point while the values of *hidden variables* (HVs) can be obtained only intermittently, at certain points of space and time. The following examples should clarify this issue:

- *Atmospheric conditions.* The OVs are measured continuously at the Earth's surface whereas high-altitude variables, such as stratospheric wind velocity and air temperature, are HVs and can be observed only when, for example, balloon-borne instruments and a radio transmitter are sent up to the "right" altitude and location.
- *Material testing.* Inexpensive and non-destructive testing of some material or product can be performed at any desired time but costly or destructive tests are carried out only sparingly.
- *Oil and mineral exploration.* The evaluation of satellite photographs, seismic experiments, geological surface studies and deep-drill work can produce OVs or HVs, depending on the associated factors of cost and difficulty.
- *Earthquake prediction.* In addition to seismographical data, a number of different OVs have been proposed and used that are likely to be correlated with the near-future occurrence of earthquakes. The HVs can be, for example, the location of the epicenter and the intensity of an earthquake to come.
- *The training of specialists.* The (controlled) OVs may be the length and intensity of training or the frequency of given exercises, while the level of actual performance under certain rare conditions represents an HV.

There is an obvious need to estimate the values of HVs for the points of space and time at which they cannot be directly and inexpensively measured. One can consider, for example, an *expert subsystem* that could be 'plugged into' an expert system in need of numerical estimates of momentarily unobservable values. We have wanted to develop such a noise-tolerant, pattern-directed inference making system that is capable of doing a relatively inexpensive estimation/prediction of HV values on the basis of inductive generalizations.

The underlying assumption is that certain mathematical properties of the behavior of *particular* OVs, on one hand, and values of *particular* HVs, on the other, are stochastically and causally related. Either can be the cause or the effect. In fact, a given OV may be the cause in some relation and the effect in some other. It is normally useful to assume that such a causal relation exists through a space or time variable representing certain constraints. We call this 'causal mediator' the *lag variable*, for reasons described below. The lag variable can assume continuous or discrete values. It can also be a counter of events, when it is time-related, or a counter of 'landmarks', when it is space-related.

A couple of examples of the physical meaning of lag variables follow. Let the OV be the movement of a mechanical device, which sends out some electrical signal. The latter causes the occurrence of an HV event, say, an underground explosion. The lag variable is now the time required for the propagation of the electrical signal. Or else, the causative OV is the use of a certain component of a car, and the effect HV is metal fatigue. The space-related lag variable represents the distance travelled by the car.

Since, as stated before, both OVs and HVs can be either cause or effect, it seemed reasonable to adopt the convention that the lag variable assumes positive values in the stochastic relations when the OV is the cause and negative values when it is the effect.

The expected result of the GPRS would enable other programming systems, such as the QO and the AT/I, to identify and estimate the values of momentarily unobservable situational variables, and the hidden reasons and consequences of decisions. Programs of this type are in general of value for *interpreting experimental data* and for *inductively detecting patterns* which underlie the rules governing a given body of observations.

5.2. The Approach

The GPRS will establish, verify and optimize stochastic and (assumedly) causal relations. As said above, these relations connect certain mathematical properties of the OV behavior and measured values of HVs in the form of generalized production rules. Before we discuss this issue in detail, we have to make a few definitions.

5.2.1. *Morphs and the Morph-Fitting Program*

Regression, Fourier and time-series analyses are the three best known and most widely used techniques to describe (that is, to explain, interpret and predict) the behavior of some 'dependent variable' in terms of others, the 'independent variables', which are normally assumed to be known exactly. There are, however, certain task environments for which even the combined and extended mathematical machinery of these powerful methods (e.g., spline functions) may not be sufficient. Such is our case in which certain other types of functional forms may have to be employed.

The mathematical properties of the OV behavior noted before refers to a set of parametric values that characterize the sequence of *patterns* of OV distribution. The patterns are approximated by one or several *basic functions.* A basic function is either a *morph* or a *delay period.* The latter represents a range of lag variable values over which the mathematical description of the OV behavior is not possible at the desired level of statistical significance and the pattern remains "unexplained" over the range in question — in simple terms, the datapoints are "too scattered".

After some deliberation, we have decided to restrict the concept of morph to three types (see Fig. 5.2.1):

- A *trend* is a straight line over a certain range of lag variable x. It is completely determined by three parameters; for example, the starting and final value of the OV, v_s and v_f, and the range of validity of the trend, d.
- A *sudden change* is a peak in the OV value with having only a momentary effect. It is determined by its location on the lag variable axis and the size of the peak.
- A *step function* is an instantaneous increase or decrease in the OV value of a lasting nature. It is also determined by two parameters, its location on the lag variable axis and the size of the change.

A couple of comments are in order. In case of two adjacent trends, the final point of the first coincides with the starting point of the second.

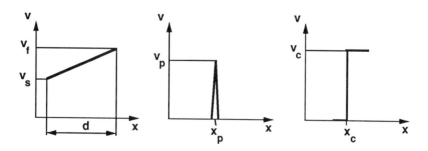

Fig. 5.2.1. — The three types of morphs: the *trend*, the *sudden change* and the *step function*. The first morph is determined by three parameters and the latter two morphs by two parameters. (A *delay period*, not shown here, is characterized by two parameters, the starting and final values of the lag variable.)

Further, the user can specify the maximum range of validity of sudden changes over the lag variable axis and thus distinguish the occurrence of a sudden change from two adjacent step functions of (almost) the same size but opposite directions.

The morph-fitting program fits a *minimum number of morphs* (and, if necessary, also identifies delay periods) for a given set of OV datapoints with the restriction that the amount of 'unexplained variance' — the sum of squares of the differences between the morph values and the OV values — is below a user-specified level of tolerance. An exemplary distribution of datapoints and morphs fitted to them are shown in Fig. 5.2.2.

An interesting by-product of this investigation was that the program could calculate in a statistically rigorous way the boundary point between two adjacent trends. This had been an outstanding problem in computational statistics; namely, where the end points of subsequent segmented (multiphase) regression lines should be.

The details of the technique and the relevant algorithm can be found in [79]. For our current purposes, it suffices to say that we can now characterize the behavior of an OV over a given range by an ordered list of a minimum number of parameter sets, each describing a basic function (a morph or a delay period) fitted to the OV datapoints.

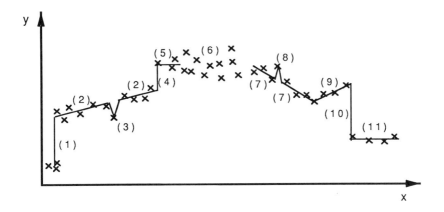

Fig. 5.2.2. — An exemplary distribution of datapoints and the morphs fitted to them: (2), (5), (7), (9) and (11) are *trends*; (1), (4) and (10) are *step functions*; (3) and (8) are *sudden changes*; the points in (6) cover a *delay period* over which the mathematical description of the OV behavior is not possible with the required level of statistical significance

5.2.2. *The Knowledge Base*

The primary constituent of the knowledge base is a set of ordered generalized production rules ('rules' for short in the following). The form of the r-th rule is

$$W_r \ / \ M_{ijk} \ / \ T_{jm} \to V_m \ (H_n) \qquad : \ Q_r$$

The three independent components on the left hand side (the 'predictor' part) are:

- W_r is the number of similar enough rules merged to form the r-th rule (see the purpose of merging later);
- M_{ijk} is the i-th combination of the parameters of the j-th morph describing the behavior of the k-th OV. Note that there can be seven possible combinations of the three parameters of a trend, $i=1,...7$ (one parameter three ways, two parameters also three ways, and all the three parameters one way); and there can be three possible combinations of the two

parameters of a sudden change or a step function, $i=1,...3$ (one parameter two ways, and both parameters one way);

- T_{jm} is the difference in lag variable value between the start of a morph and the occurrence of the m-th value of the HV considered, H_n. T_{jm} is positive when the OV is the cause (the start of a morph precedes the HV occurrence), and negative when OV is the effect (the occurrence of the HV value precedes the start of the morph).

On the right hand side (the 'predicted part'), we have

- V_m, the m-th value of H_n, the n-th HV;
- Q_r is the quality or the 'credibility level' of the r-th rule. Its value is between 0 and 1, and depends on
  - how well the morph in question fits the datapoints over its range of validity, and
  - how many and how similar are the rules merged into the r-th rule.

(Some additional information is also attached to each rule, such as the number of datapoints fitted by the morph and the results of certain statistical computations — for example, the numerator and the denominator of the relevant F-ratio.)

The above implies that the program has to adjust the credibility level of every rule after the rule merging operation. (More will be said about it in the next section.) It should be noted that since the delay period represents the lack of reliable information about the OV behavior, it is not used in the formation of rules. Furthermore, a particular OV may be referenced in the left hand side of several rules; a particular morph can be represented in several different parameter combinations vis-à-vis the same HV; and in turn, the same HV value can be associated with several different OVs and through several different morph parameter combinations.

The following gives a concise overview of the entities introduced:

knowledge base ::=   <ordered set of merged rules>;

merged rule ::=   <number of similar rules merged>,
                            <weighted average of rule parameters in
                            merged rules>, <modified credibility level>;

similar rules ::=   <rules with rule parameters in predefined
                            proximity>;

rule parameters ::=   <morph parameters>, <lag>,
                                  <value and type of HV>;

morph parameters ::=   <type of OV>, <trend parameters> |
                                  <step function parameters> |
                                  <sudden change parameters>;

trend parameters ::=   <{starting value, final value, range}>;

step function parameters ::= <{location, size of change}>;

sudden change parameters ::= <{location, size of peak}>;

basic pattern ::=   <morph> | <delay period>.

The user can significantly reduce the possibility of a combinatorial
explosion; that is, the case of unmanageably large search spaces. After a
sufficient amount of data of OV and HV values are read in the computer, he is
asked by the system about

- which are the OV—HV pairs that are likely to be causally related;
- which is the cause and which is the effect in a given OV—HV combination
  (to determine the sign of the lag value and to disable the forming of
  incorrect causal relations);
- the maximum and minimum meaningful values of lag in a given OH—HV
  combination (likely limits of *causal relevance*).

The more information — in terms of answers to the above questions — the
user is able to provide, the less the number of meaningless rules established
will be. The system will then produce the initial knowledge base containing
all the rules that satisfy the above restrictions. However, this initial
knowledge base normally contains many rules that are due to some
coincidence rather than describing actual *causal relations*. A powerful
learning facility, to be discussed in the next section, will "clean up" and
optimize the knowledge base.

5.2.3. *The Optimization of the Knowledge Base*

Let us consider a "true" causal relation. There is a *population* of an indefinite
number of observable rules that represent that causal relation. The

parameters of these rules have certain probability distributions around the *true value* — due to statistical fluctuations, measurement errors, variations in the environmental conditions, etc. Each time a rule is established in the knowledge base, it is a *sample* of that population. The more such similar rule samples the system identifies and merges, the better estimates of the true rule parameters can be obtained and, consequently, the quality (credibility) of the resulting rule increases. In other words, the higher number of merged rules are represented by a particular rule, the more evidence the system has that the rule in question corresponds to some causal relation and is not due to a statistical coincidence. The rules in the knowledge base are ordered according to decreasing values of credibility. After each merge operation — affecting a set of rules which are similar enough according to the user's specifications — both the credibility value and the position of the new rule in the linear hierarchy has to be computed.

Obviously, two rules can be merged only if

- both rule relate the same OV—HV pair and the respective lag values are of the same sign;
- the same type of morph appears in both rules;
- the same combination of morph parameters appears in both rules;
- the corresponding rule parameters are within the allowed (user-specified) range from each other.

We have described in [64] the computational method for deciding as to whether to merge two or more rules, and how to update the credibility values, F-ratios and other parameters. In view of the fact that experimental data about the OV and HV values may become available continually and at irregular times, the process of optimizing the knowledge base has an intermittent mode of operation, similarly to garbage collection with list processing languages.

5.2.4. *The Estimation of Values of Hidden Variables*

The program, as originally developed, could yield *point estimates* of HV values. We describe this first, before discussing the extended ability of computing *functional estimates* [71].

Suppose the user needs a point estimate of an OV value at a given point on the lag variable scale, say, at x_i. He has to provide OV datapoints that either precede x_i (when OV is the cause) or succeed it (when OV is the effect). The range of the usable OV datapoints must not be closer to x_i than the minimum and farther than the maximum meaningful lag value difference, as specified by the user at the beginning of data input to the system.

The morph-fitting program fits new "predictor" morphs through these datapoints the same way it was done at the time the knowledge base was first established. The system then tries to match the resulting morph parameters, called the *query*, with the left hand side of the most credible rule in the knowledge base. A matching rule must

- connect the same OV—HV pair;
- refer to the same morph type;
- involve morph parameter and lag variable values "similar enough" to those in the query — that is, within the user-specified range of merging rules.

The user may request the N best estimates at the point x_i. Since the rules in the knowledge base are ordered according to decreasing values of credibility levels, it seems natural to return N values (or less if the knowledge base cannot provide enough) obtained through those rules that are as close to the top as possible and satisfy the criteria of match. However, the *overall* quality of the estimates, their so-called *confidence level*, depends on two additional factors. (The quality order of the *estimates* may, therefore, be different from the order of the *rules* through which they were obtained.)

The additional factors contributing to the quality of the estimates are

- how well the predictor morph fits the OV datapoints,
- how close the parameters of the predictor morph are to those of the matching rule in the knowledge base.

These measures are again probabilistic in nature and can be computed in a way similar to that employed in merging rules. We must state that the N best rules do not necessarily provide the N best estimates. The program has to compute more estimates; this process will be terminated when

- the N best estimates have in fact been obtained; or
- the number of rules matching is less than N; or
- the credibility level of the last rule found is less than the lowest confidence level obtained so far; or
- there are no more rules to consider.

We have used an additional heuristics. The estimates are given to the user only if the *weighted average* credibility level of the rules used for the estimation is at least 0.75. The weighting is by the rule weights, W_r, the number of rules merged in the given r-th rule. This requirement will balance highly credible rules of small weight (formed by merging few rules) and less credible rules of large weights (the resulting rule being "blurred" due to incorporating many somewhat dissimilar rules).

It is important to realize that after a sufficiently long period of collecting data, establishing rules and optimizing the knowledge base, the rules that represent "real" causal relations will tend to move toward the top, and those rules that were established by chance co-occurrence of data will percolate downward and will not contribute to the estimation process.

Next the procedure of obtaining functional estimates of HV values is described (see [71]). The program components are identical to those used for the point estimation, only a new executive had to be written for the task. The following steps are taken:

- Compute a point estimate for the left-most point of the range over which a functional estimate is needed. (This point estimate can be the best available or the average of all those whose confidence level is above a user-specified value.)
- Compute the *range of validity* of the point estimate. (An effective and statistically justified algorithm was developed for this purpose.)
- Obtain further point estimates at the end of successive ranges of validity if possible. (Whenever it is not possible because a delay period is identified instead of a "predictor" morph, there is a discontinuity in the functional estimate.)
- Each point estimate is duplicated over its range of validity a number of times that is proportional to its confidence level. This way, we have given each point estimate a weight equivalent to its credibility level.
- A non-equidistant sequence of point estimates has now been computed. These points are within a combined range of validity that covers the concatenated sequence of the respective ranges of validity.
- Invoke the morph-fitting program to fit morphs to the weighted point estimates. The combined 'goodness-of-fit' of the morphs, available from the computation, is the quality of the functional estimate. The range of validity of the functional estimate is equal to the combined range of validity of the point estimates.

Finally, we note that a screen-based and a hardcopy-oriented plotting facility enables the user to assess the quality of the result visually as well. If he is not satisfied, he can change certain parametric values in the morph-fitting program or in the rule merging algorithm in order to modify the outcome of the computations. (See, however, Section 5.4.1 with reference to the parameters of the morph-fitting program.)

5.3. System Modules

There are six distinct but interacting, major modules. These and some of the utility programs are described briefly below. (Some of the modules of GPRS are applicable on their own for various problems — similarly to the modules of the QO system.)

5.3.1. *The GPRS-1 Module*

This program component first receives prompted answers from the user to a set of questions. These specify certain parametric values and a few details about numerical procedures. (In a later version of GPRS, described in Section 5.4, the user provides only an initial set of tentative values for these parameters and a heuristic optimization method, based on *pattern search*, produces the most suitable values.)

A minimal number of morphs (and, if necessary, delay periods) are fitted to the given set of datapoints. The morphs also satisfy the restriction that only a prespecified maximum amount of *unexplained variance* (noise) is allowed around them.

5.3.2. *The GPRS-2 Module*

This module sets up *all possible* rules, as defined in Section 5.2.2, containing the predictor part, the predicted part, and certain additional information. These rules connect those OVs and HVs that are said by the user to be potentially causally related. The "causal direction" (depending on whether the OV is the assumed cause or effect), and the upper and lower limits of the lag value are also taken into consideration.

5.3.3. *The GPRS-3 Module*

This part of the system merges similar enough rules. Only those subsets of rules are considered in which

- the same OV and the same HV are connected;
- the OV is described in terms of the same type of morph and the characterizing set of morph parameters are identical;
- the rule generation was done according to identical criteria;
- the morph parameters, lag values, and the HV values are within the user-specified tolerance limits.

The module then merges the rules that satisfy the above conditions into a single rule whose parameters assume the weighted average values of the respective constituents. Further, the credibility value of the resulting rule is raised to a new, statistically rigorously determined level.

5.3.4. *The GPRS-4 Module*

This program component performs the computation for estimating/predicting an HV value at a given point on the lag scale. It looks at the knowledge base and identifies those OVs

- that are causally related to the HV in question, and
- for which a sequence of datapoints is available within the range of lag values determined by its upper and lower limits.

The user may request the *N* best estimates or just a single one which is the average of the acceptable values.

5.3.5. *The GPRS-5 Module*

This module extends the system to *distributed processing and intelligence.* Let us assume that there is a star-like computer network (see Fig. 5.3.1) in which the central machine and the processors on the periphery collect data and establish separate knowledge bases. (Such a system is appropriate, for example, in an earthquake-prone region in which data collection and prediction of undesirable events are important.)

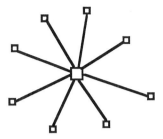

Fig. 5.3.1. — A star-like computer network. The large central processor and each of the smaller peripheral units obtain data from the environment and establish from these separate knowledge bases. The user can merge the 'raw data' files and the knowledge bases after the system verifies that certain of their known statistical properties, and the rule and file generation criteria are identical.

Naturally, the larger the knowledge base is, the more reliable are the results of prediction/estimation. The user of the network may at times wish to combine the data files and the knowledge bases in the central machine. The GPRS-5 modules performs that operation once it has verified the *compatibility criterion*, namely, that all relevant statistical properties of such records, and the respective rule and file generation criteria have been identical.

5.3.6. *The GPRS-6 Module*

Finally, this part of the program extends the system's capabilities to compute the *functional form* of HV distributions as described in Section 5.2.4. Again, several versions of the algorithm are available for the user to select from.

5.3.7. *The Utility Programs*

There are several utility programs — some of these are transparent to the user; that is, not directly invoked by him. A few of the important utility programs are:

- a facility that prompts the user for input and verifies, whenever possible, the data received;
- an input program that loads OV and HV data from tapes or disks;
- a 'self-explication facility' that traces and, upon user request, displays in a simple English-like language the rules employed in the estimation process;
- a plotting routine that provides a visually useful version of the datapoints and the morphs fitted to them, either on the screen or on a hardcopy.

5.4.   The Implementation

The GPRS is a fairly large system. Several components have to co-exist in the core memory at the same time. We have also tried to make sure that memory swapping is minimized. We describe here only two interesting aspects of the system.

5.4.1. *A Pattern Search Method for Optimization*

It was noted before that the user has to specify a certain number of parametric values for the morph-fitting program at the beginning. It turned out that even after quite some experience with the system, we could not identify a reasonable set of parametric values for a given set of datapoints on the first try and had to go through several steps of iteration before we were satisfied with the results of morph-fitting. We have therefore decided to try to automate and optimize the process of finding parametric values. This program has turned out to be a rather general-purpose method, applicable also in other areas of optimization.

There is a frequent need to optimize multi-dimensional, implicitly specified, 'non-smooth' functions — that is, those with ridges, plateaus, discontinuities and the like. Analytical and traditional approximation techniques cannot be used, and heuristic methods have to be employed.

We have developed an inexpensive but effective version of 'pattern search' [74]. It is based on the heuristic of repeating the pattern of the best previous search direction, found in exploratory moves, as long as it is advantageous; that is, as long as the value of the *response* — the *objective function* — improves. (The response value in the case of the GPRS is the quality of fit by the morphs, and the variables are the parameters needed for the morph-fitting program.)

Pattern search in general refers to a systematic *direct search* method. First, *exploratory moves* (going from one point to another in the N-dimensional space of the N independent variables) provide some basic knowledge about the behavior of the objective function. So-called *pattern moves* are then made, which repeat the best exploratory move leading to the current base point. The step size is increased when there is 'success', the objective function returns a higher value, and the step size is decreased when a 'failure' is encountered.

In general, the method needs a number of starting points, one with each of the first exploratory moves. This way, the program does not get trapped in a local optimum. We have therefore developed a systematic search algorithm to generate first a sequence of well distributed starting points. We have proved empirically that the final results (i.e., the location of the optimum response values found) are superior to those when randomly generated starting points are used. The comparisons were based on

- various dispersion measures;
- the number of starting points followed by pattern moves converging to the solution (as opposed to *not converging* pattern moves);
- the number of iterations needed before the solution is reached (the distance between the starting point and the solution).

We have also introduced a heuristic for the termination of the search process. No further starting point is generated when the *largest* maximum value of the objective function and its location have not changed over the last 10 different starting point values. Full and partial tracing modes enable the user to monitor the exploratory and pattern search processes at various levels of detail.

Finally, it should be noted that direct search methods seem to have distinct advantages over other techniques:

- there is no need to compute the values of derivatives;
- the search process is robust and converges rather rapidly to the solution due to simple and plausible heuristics;
- the whole approach is effective by virtue of the pattern moves used, the boundaries identified for the domain of optimization and the judicious halting criterion employed in the search for a global optimum.

5.4.2. *The Interaction with the System*

The second aspect of the implementation to be mentioned is with reference to the high level of interaction between the user and the system. The system guides and, whenever possible, verifies the user actions. After signing on for the GPRS, the user is shown on the screen a menu from which he is to select one item at a time. The options are:

(1) *Introduction* provides a short description of the objectives and the user environment of the GPRS.

(2) *User's Manual* contains a detailed discussion of how to input values for the OV and HV datapoints, how to respond to system requests, how to initiate the morph-fitting program, how to obtain point and functional estimates of the HV behavior, how to get the system to "explain itself" (that is, to display in English the rules employed in the estimation process), how to inspect raw data files, how to merge raw data files and knowledge bases (which may originate from geographically separated processors), etc.

(3) *Load OV Data* enables the user to input values for OV datapoints, to invoke the morph-fitting program, and to specify the limits of meaningful lag values.

(4) *Load HV Data* enables the user to input values for observed HV datapoints, to specify potentially causally related OV—HV pairs, to indicate which is the cause and which is the effect, to have rules automatically established, merged and displayed, to eliminate any rule not wanted, and to update the knowledge base.

(5) *Prediction* yields a high level guidance to accomplish point and functional estimations/predictions, offering numerous user options in the process.

(6) *Display in English the Rules Employed* traces the system's reasoning process during the prediction activity.

(7) *Merge Designated Knowledge Bases* performs this action after the system verifies that the conditions for it are satisfied.

(8) *Merge Designated Raw Data Files* again performs this action after the system verifies that the conditions for it are satisfied.

(9) *Inquiring about Raw Data Files* causes the system to display potentially causally related OV—HV pairs, the corresponding limits of meaningful lag variable values, the minimum degree of similarity between mergeable rules, and the like.

(10) *Exit.*

5.5. Applications

We have applied the GPRS in two areas. Although the first application was in the domain of macro-economics, the primary objective was to test the flexibility and power of the whole system, and to see the benefits of the pattern search technique. The first eight variables shown in Table 5.5 were considered OVs and the last variable was the HV.

Table 5.5.1. — Nine macro-economic datasets; the first eight represent OVs and the last one is a HV that needs to be estimated at arbitrarily chosen time points

OV	Variable Name	Records Available From	To	Frequency of Data
1	Foreign Trade Balance of U.S.	1934	1979	yearly
2	Consumer Price Index	1934	1981	yearly
3	Business Failure Rate	1930	1980	yearly
4	Prime Rate Charged by Banks	1975	1981	monthly
5	Average Gold Price	1975	1981	monthly
6	Average Price of Crude Oil	1978	1981	monthly
7	Average Wholesale Price of Gas	1978	1981	monthly
8	Average Price of Silver	1973	1981	monthly
9	Unemployment Rate	1966	1981	monthly

We have wanted to investigate one of the learning aspects of GPRS. Thirty HV datapoints, in two groups of 15 each, were selected at random out of the 184 available monthly figures — covering the period from January 1966 to April 1981. The unused 154 values served for an objective comparison between predicted and actual values. Without going into the details of the econometric results and their statistical significance, we note that the OVs 3, 4, 7, and 8 were found to be in a causal relation with the HV. Interestingly enough, each of the OVs can be both cause and effect. Further, there was an important improvement in the estimation process when 30, instead of 15, HV values were used in constructing the knowledge base. The merging of similar rules has also resulted in better and statistically more significant estimated values. Finally, the parameter optimization module was found to be a cost-effective, general-purpose tool. It noticeably increased the accuracy of the estimated values.

The second application was with reference to the man-machine interaction in a simulated Air Traffic Control environment (see Chapter 8). The details of the general problem area will be discussed later. We only note here that the role of GPRS was to estimate the values of situational variables, the type of decisions and their underlying reasons when they were not directly observable.

5.6. Summary

GPRS is a noise-tolerant, pattern-directed inference system that is capable of making inductive generalizations in estimating/predicting unknown values of HVs — variables that can in general be observed and measured only intermittently, at irregularly distributed points of space and time. The result can be either the point or the functional estimate of the behavior of such variables.

The implementation is a strongly-coupled interactive system which makes full use of the user's knowledge about causal relations and the constraints they must satisfy. In relying on this knowledge, the system reduces the danger of combinatorial explosion. One of the user options extends the applicability of GPRS to distributed processing and intelligence.

The system can be used as an *expert subsystem*, a computational module of an expert system in need of numerical estimates of values of variables that cannot be measured at arbitrary points of space and time.

5.7. Acknowledgements

Ron Lo has contributed to all facets of this project. Ernesto Morgado has worked on the morph-fitting program and Cher Lo on the pattern search component. John E. Brown and Han Yong You have also participated in the development of the GPRS.

6. Distributed Planning and Problem Solving Systems (DPPSS)

6.1. Introduction and Research Objectives

In continuing the ideas presented in Chapter 2, it will helpful if we consider plans as representations of proposed courses of action and of the agents to carry out the actions. Human or machine decision makers have to reason about and evaluate plans *before* and *during* plan execution.

Let us assume that there is a finite set of primitive actions,

$$A: \{a_1, a_2, ..., a_m\}$$

Each of these is capable of being carried out if

- certain preconditions are satisfied, and if
- it receives the necessary input (information, "material", starting time wanted).

The outcome of the operation is successful if

- certain postconditions are satisfied, and
- produces the expected output (information, "material", satisfactory finishing time).

There are also a given number of processors,

$$P: \{p_1, p_2, ..., p_n\}$$

Each processor is characterized by

- the subset of the total set of primitive actions it can perform, and
- at what cost and
- with what speed.

(The value of the latter two can, in general, have certain probability distributions.) Given are then the current status of the environment, including that of the processors, and a desired change in it — for example, a new spatial arrangement of existing objects or new objects to be produced. (A well-known example of the traditional approach to planning is the STRIPS system [56].) The role of planning is to answer two basic questions:

- what (weakly or partially) ordered sequence of primitive actions can produce the outcome wanted, and
- how and in what order should the primitive actions be assigned to the individual processors, with the constraint that
    ·minimum overall production time or
    ·minimum overall production costs are to be achieved.

As one can see, the original AI problem domain of planning is extended by the optimization requirements and now also involves the concerns of operations research and decision theory [113]. The above paradigm applies to a wide variety of activity, ranging from robot planning and natural language generation to automatic programming and experimentation in fields such as molecular genetics, chemical synthesis, agriculture and animal breeding.

Realistic planning must also take into consideration the fact that it is not possible to have a complete knowledge of the relevant parts of the world (its current status, the effects of actions, etc.) and that the available knowledge may well be inconsistent. It is usually not feasible to perform the planning operation so that we branch into 'all possible worlds' at the points of uncertainty because of the obvious danger of a combinatorial explosion. Under certain circumstances, the approach of interleaving plan formation and execution may be effective. (Plans are generated into a certain depth of time relying on partial information and then are continually modified when the actual and the expected worlds become "too" different).

6.1.1. *Traditional AI Approaches to Centralized Planning*

The following, not completely distinct, approaches have been followed by AI researchers:

- *Hierarchical planning* constructs a hierarchy of plan representations, each level being associated with a different degree of plan abstraction. It means that the lower the level in the hierarchy the planner is at, the more detail is considered for the actions.

- *Non-hierarchical planning* is done at one single level with all possible details considered. The undesirable consequence may be that planning consumes too much computing time and loses its cost-effectiveness.
- *Script-based planning* relies on stored 'plan skeletons' ranging from very general ones (to be used first) to rather specific ones (to be used for details toward the end of the planning process). The plan skeletons are filled in with task-specific information using one of the other methods being discussed.
- *Opportunistic planning* is *asynchronous* (namely, different subplan specialists, independently from each other, put suggestions on and read necessary information from a common information structure called a 'blackboard'). It is also *oppportunistic* (planning decisions are made only when the need for them arises).
- *Simulation-based planning* relies on look-ahead simulations in an extrapolated world, yielding desirable plan segments at decision points.

The two main problems that arise with centralized planning are:

- how to *limit the search* for an optimum (or, at least, satisfactory) ordering of primitive problem solving actions without getting into a combinatorially explosive situation, and
- how to *resolve the conflict* between interacting subproblems and subgoals when the preconditions of some actions interfere with each other. (Premature commitment to an arbitrary combination of interacting subproblems may necessitate lots of backtracking and additional search.)

6.1.2. *On Distributed Planning and Problem Solving*

There are problem-solving tasks whose size and certain other characteristics do not allow them to be processed effectively and efficiently by a single computer system. Such tasks are characterized by one or several of the following properties:

- spatially distributed input/output (or sensor/effector) operations,
- a need for extensive communication over long distances,
- time-stressed demands for solution,
- a large degree of functional specialization between quasi-simultaneous processes,
- a need for reliable computation under uncertainty — that is, in relying on incomplete and inconsistent data and knowledge,

- a need for graceful degradation — the results are acceptable (although possibly of lower quality and obtained more slowly) even when some of the computational and communicational facilities have become disabled,
- a limited amount of shared resources must work on common objectives, without competition and conflict of interests.

Advances in computer and communication technology have made studies in *Distributed Artificial Intelligence* (DAI) possible. The area of our concern, *Distributed Planning and Problem Solving Systems* (DPPSS) is the most active one in DAI. DPPSS is considered as the combination of AI and Distributed Processing methodology. Let us look at the relevant concepts first.

Distributed Processing is characterized by the following:

- several dissimilar tasks are being solved simultaneously by a network of processors,
- the management of common access by the individual processors to a set of resources is the main reason for the interaction between processors,
- resource distribution and conflicts between tasks are hidden from the user,
- from the user's point of view, each task appears to be performed by a system dedicated to that task only.

The hardware used in Distributed Processing is readily available for DPPSS. The underlying philosophy of such networks is *cooperation* — rather than competition as found in biology, sociology and various organizational systems. The allocation of network components to problem solving processes is explicit and deterministic but may change as the environment changes. The interactions between the network components and the decisions made by them can be known to the user if necessary. Each node of the network possesses enough problem solving knowledge to apply its own expertise to its task and to communicate with other nodes. The problem solving strategy of every node is in harmony with that of the others.

Our research objectives have been set as follows.

- To study how a group of agents, a *network of processor units*
  - *cooperate* with a selected set of others to achieve a common set of objectives,
  - are *interconnected* for full utilization of resources and effective goal accomplishment (network architecture),
  - share *distributed knowledge* and handle *uncertainty*,

- *perform* as a function of the amount of knowledge and meta-knowledge[1] available to the individual nodes,
- can *reconfigure the network* in response to a dynamically changing environment;
- To study *individual agents' planning activity*, and the influence of different architectures on the processes of *communication and control*;
- To *apply the results* to a set of different domains, each with different reliability concerns, quality measures, timing aspects, and computational and communications requirements.

We have hoped that we would arrive at a *unified conceptual framework* for plan generation by distributed processors, and the conclusions drawn from our studies would not only enhance the *automation* of several technologically and economically important activities but may also shed light on how human individuals and groups *organize their knowledge, coordinate their actions*, and *interact with each other.*

6.1.3. *General Categories of Application of DPPSS*

There are four basic but overlapping categories of *applications* of DPPSS:

- Through integration and analysis, one tries to *interpret* distributed data and form a *model* of its generation, which should help in its use. Examples of such a need can be found with network fault diagnosis or with distributed sensors.
- *Distributed planning and control* identifies and coordinates actions to be performed by a set of distributed effector nodes. Such operation is found, for example, with cooperating robots, remotely piloted fleets of drones, intelligent command and control networks, and in the four domains in which we have been working (see Sections 6.2 - 6.5).
- *Coordination networks* enable, for example, multi-participant project coordination and cooperation among workstations working on common objectives.
- *Cooperative interaction among expert systems* may aim at, for example, diagnoses on patients suffering from diseases of different categories.

---

[1] Meta-knowledge references knowledge *about* knowledge. It is often necessary to know where additional knowledge is available from, how reliable and expensive certain knowledge components are, whether new knowledge should be separated from the existing body of knowledge, what properties of a certain piece of knowledge are to be kept and what is the intended use of knowledge.

6.1.4. *The Four Phases of Network Activity within DPPSS*

The problem solving process goes through the following phases recursively:

(1) *Problem decomposition* into subproblems. The network may be viewed either in a 'reductionist' manner (the objective is to pull apart a centralized system into distinct components) or in a 'constructivist' manner (the objective is to organize a set of subsystems into a society of cooperating nodes).

(2) *Subproblem distribution* among qualified nodes can be

- *functional* — each node is an expert in some part of the problem domain and subproblems are routed to the appropriate expert;
- *spatial* — each node is a "universal expert" and selects that part of the whole problem that is the nearest to it geographically;
- *hierarchical* — corresponding information and control of different levels are concentrated in identical points of the network, which makes the problem solving process rather sensitive to the loss of a high-level node;
- *heterarchical* — cooperation between nodes at one level makes it robust to loss of nodes but suffers from increased problems of communication and control.
- *redundant* — makes use of alternative views of data and tasks but is less efficient and requires more network resources.
- *disjoint activities* — the lack of redundancy may lead to degraded performance when a relevant and crucial piece of information is lost.

(3) *Subproblem solution* still cannot be viewed as an independent node activity because of the concerns with timing and coherence (see later). The local interpretation of data is spatially and temporally interdependent.

(4) *Answer synthesis* also involves the elimination of incomplete and inconsistent solution components. There is an increased emphasis at this level on the timeliness and the reliability of the end product.

6.1.5. *A Few Major Issues with DPPSS*

One has to be concerned with the following major issues:

- The *connection problem* is about task allocation, how nodes with tasks find nodes capable of executing them. For example, in the 'contract net' approach to be discussed later, the negotiation process involves 'announce-bid-award' cycles of interaction.
- The *limited communication-bandwidth problem* affects reliability, cost and delays. The trend indicates that computation is becoming cheaper and more reliable much more rapidly than communication. Therefore, communication is the primary candidate for minimization in order to optimize overall performance. There are three modes of communication: broadcasting at large which aims at all nodes, group-broadcasting to a subset of nodes selected on the basis of 'need-to-know', and point-to-point messages with specified addresses.
- The *coherence problem* references the requirement that the actions of the individual nodes be mutually supportive and not interfering. They rely on hierarchically distributed knowledge and meta-knowledge to balance the load among the nodes, perform *network perception* and the necessary communication for *global coherence*. (Our relevant "perceive-plan-act" cycle is introduced later.)
- The *timing problem* concerns the fact that subtasks must be executed not according to the "first-in-first-out" discipline but according to their level of urgency and in consideration of the needs of other tasks.
- The *problem of network architecture* has several possible solutions in interconnecting individual processors as
  - *master-slaves*, characterized by forced cooperation and fixed relations,
  - *same-class-citizens*, characterized by free cooperation and "friendly" negotiations,
  - *self-reliant units* with no cooperation and minimum communication, to be used only under exceptional conditions.

It should be noted that any node in the network may appear in several different roles vis-à-vis the others. The control architecture should be such that it is *generated* and, if necessary in response to a change in the environment, *reconfigured dynamically* during the problem solving process. It is desirable that only a *part* of the whole network should be affected by such reconfiguration. An attractive approach to such *dynamic plan mending* makes use the so-called dependency-directed backtracking and tries to make a minimum amount of change to the existing plan.
- The *problems of learning and sharing improved skills* utilizes a knowledge base containing past problems and their solutions. The solution to a new

problem starts with the solution to a 'similar' problem encountered before, which is then modified according to some heuristically guided technique.

- The *problem of coordination* may be effected by systematic signalling between the different planning activities. (Cf. the *distributed scratch pads* used in our project on Distributed Air Traffic Control.)
- The *problem of uncertainty* (incorrect or incomplete knowledge) arises because of unexpected changes in the environment, inherent noise and delays in communication, unforeseen consequences of actions, and "mismatches" between problems and solution — for example, when a functionally distributed solution does not suit a spatial task distribution. The uncertainties to be dealt with can be
  - *environmental* — the status of other processors and communication channels,
  - *data-oriented* — incomplete and inconsistent local data at a node,
  - *control-oriented* — inaccurate model of the other nodes' activity.

Continual feedback from the environment should serve as an indication of the need for replanning.

6.1.6. *Other Dimensions of Classification of Distributed Planning*

The division of work among the processors is relevant from the viewpoint of reliability and cost-effectiveness. It can be classified as follows.

- *Individual planning* by each processor in a manner that is
  - *object-oriented and autonomous* — the nodes are self-reliant; the interaction between them is only through sensing (e.g., by associated radar equipment in the case of air traffic control);
  - *object-oriented and cooperative* — the interaction between the nodes is through both sensing and negotiations; they can also adhere to the "rules of performance" (e.g., the rules of traffic in controlling street traffic).
- *Group planning* in a manner that is
  - *location-centered* — one selected processor performs the planning for its spatially clustered group (e.g., all manufacturing plants within a region) and interacts with the selected processors of each other group affected by the plan;
  - *function-centered* — one selected processor performs the planning for its group of members with identical function (e.g., all airplanes in the

take-off phase) and interacts with the selected processors of each other group affected by the plan.
- *Global Planning* in a manner that is
  - *segment-centered* — qualified processors perform a segment of the planning; a segment may refer to a location, a function or node type.
  - *hierarchically divided* — selected processors are in charge of low, medium and high level planning, each with sharply defined authority.

The related research activity can be viewed in three overlapping domains:

- Network architecture to contribute to
  - the design of individual nodes,
  - the design of the interconnection mechanism.
- System design to enhance
  - message delivery and interpretation,
  - knowledge base consistency over the whole network,
  - knowledge base recovery when uncertainty arises.
- Problem solving methodology to take care of
  - inter-node control (authority allocation),
  - knowledge organization (partial knowledge distribution).

One must keep in mind that often it is impossible to know the state of the world completely, the consequences of actions cannot be predicted with certainty, and a recovery from the effects of inaccuracy cannot be achieved during plan execution.

6.1.7. *Global Coherence with Decentralized Control*

A coherent problem solving network with decentralized control needs

- *coverage* — every necessary portion of the overall problem description must be included in the activities of at least one node,
- *connectivity* — the nodes must interact so that coverage can be utilized and the overall solution can be arrived at,
- *capability* — the communicational and computational capabilities of the network should suffice to achieve coverage and connectivity.

It is easy to see that, in general, "global control" cannot be assigned to a single node because of the limited internode communication which restricts the view of the other nodes. The global control arrangement would lead to a severe communicational and computational bottleneck, and the desirable network property of graceful degradation would become infeasible. Each node must, therefore, direct its own activity in concert with others, and rely on incomplete, inaccurate and inconsistent information.

6.1.8. *The Issue of Time-Criticality*

We call planning (and the related scheduling) *time-critical* when the time scale of satisfying various constraints is commeasurable with the time requirements of the planning process itself. The time available for responding to anticipated events varies due to changes in the environment and, even more importantly, unexpected events may occur and generate a need for replanning in limited time. Often certain commitments can be made only when some information becomes available and it is not known in advance when it happens.

The traditional approach in computer science, emphasizing the concerns about complexity and correctness, is unsatisfactory since getting a least complex and optimum solution late is as good as getting no solution at all. A few of the techniques discussed in the literature is briefly described next.

Michalski and Winston [153] proposed a variable precision system in which the term *precision* can refer to specificity or certainty. Namely, the solution may become more specific and/or certain if the system is given more time. To implement this idea, they introduced so-called *censored production rules* of the form

```
if <premise>
then <conclusion>
unless [<censor>]*
```

Censors can thus be interpreted as an exception to the rule,

$$\text{<premise>} \land \neg [\text{<censor>}]^* \Rightarrow \text{<decision>}$$

but of lesser importance (low likelihood assertions). If there is insufficient time or resources to determine the censors' truth value, some or all the censors may be ignored. The authors suggest that, ideally speaking, rule premises and censors should be associated with numerical estimates of the likelihood of their being true and of the related testing costs.

Another suggestion by Dean and Boddy [19, 42] is to suspend a planning procedure when necessary and resume it when time becomes available. The result obtained "anytime" latest should be used if needed. Our approach is somewhat related to this idea and is described next.

On-line systems in general, such as some expert systems and planning systems, may have time-criticality at a *high-level* (for example, in the computerized intensive care unit of a hospital or in an air defense environment), *medium-level* (for example, our air traffic control and street traffic light control systems) or *low level* (for example, our distributed manufacturing control system). The approach we have developed with the latter two categories can be termed *iterative constraint satisfaction*. (The word 'constraint' is used in the generic sense and can also mean 'requirement' or 'condition'.) A specially designed program module, the *Time-Criticality Supervisor* oversees, interrupts if necessary, and coordinates any part of the planning process, communicates with the changing task environment, updates the relevant part of the knowledge base and, in general, tries to make sure that the plan execution is near-optimal.

The constraints are ordered according to the *priority* of each to be satisfied by the computational results. A priority level is the product of two user-specified measures, the *importance* and the *urgency* of the constraint in question[1]. In view of the non-deterministic environments, there is no way to predict the length of the calculations that would satisfy one, several or all constraints. Appropriately sized time slices are allocated to the computation, and the program attempts to satisfy one constraint after the other. At the end of every time slice, the Time-Criticality Supervisor decides whether the next time slice is to be used for refining the results further or the so far best result has to be already returned for use.

6.1.9. *Two Successful Approaches to DPPSS*

We now discuss briefly two approaches that have proved to be successful in arriving at an effective solution to the above problems.

---

[1]The *importance* of a constraint is task-dependent and remains constant during planning generation and plan execution. *Urgency*, on the other hand, is determined by the difference between current time and the latest possible time by which the constraint has to be satisfied. The product of the two, the *priority* level of the constraint, balances the two considerations.

In some task environment, more than two factors may be relevant. For example, in an anti-missile defense system, the time difference between the expected time of impact and the current time determines the urgency level whereas the estimated destructive power of the missile and the importance of the expected target represent two additional factors contributing to the priority level.

The *contract network* by Davis and Smith [41] consists of entities willing to negotiate to obtain coherent behavior. This involves multi-directional exchanges of information between interested parties, the evaluation of the transmitted information by each member from its own perspective, and a final agreement by the mutual selection of planning tasks. A network of contracted control relationship is established between the nodes. The contracts, through which the nodes coordinate their activity, are elaborated in a top-down manner. At each stage, a manager node decomposes its contract into subcontracts to be accomplished by other nodes. This procedure continues until each node at the bottom level can complete its contract without assistance. The resulting manager-contractor relationships are distributed throughout the whole network. During the answer synthesis, the solutions to the individual subproblems are collected and put together by the respective managers.

The other approach is called *coordination using organizational structures* (see, e.g., [35]). Similarly to the contract net concept, it recognizes the individual nodes' limited view of the network problem solving activity, and that a network coordination policy must not consume more resources than the benefits obtained through the increased coherence in problem solving. The approach aims at an acceptable, rather than optimum, network performance. The organizational structure introduced specifies a general set of node responsibilities and node interaction patterns available to all nodes. The control decisions are long-lasting, and represent a balance between a precise local-node-control for planning sequences of activities and a network-node-control for the construction and maintenance of a network-wide organizational structure.

6.2. A Distributed System for Air Traffic Control

6.2.1. *General Issues*

The task of Air Traffic Control (ATC) is highly expert-labor-intensive, and requires considerable intelligence to perceive the essential aspects of situations and to make satisfactory decisions under time-stressed conditions. To increase the productivity of ATC, significant research efforts have been made within the U.S. Federal Aviation Administration and at different research institutions on how to delegate more responsibility to computers. Our work, at a more modest level, has attempted to contribute to the understanding of the ATC processes, to provide practical aid for the training and evaluation of human controllers (see Chapter 8), and to make some steps toward future automated ATC systems.

The ATC task environment is especially suitable for AI studies because

- the intellectual component is complex for human operators and the economic importance is extremely high;
- the air traffic situation in many parts of the world is critically saturated and current technology is unable to satisfy near-future needs;
- one can identify problem areas of different sizes that can be attacked successively, and automated segments of the whole task can co-exist with task segments performed by traditional, "manual" methods;
- until an automated subsystem is fully developed and tested, it can function in a realistic, simulated world;
- an automated system would share many challenging properties with other complex systems, such as short and long term planning, time-stressed decision making under uncertainty and risk, problem solving in dynamically changing domains in which the solutions satisfy a hierarchy of constraints;
- one can define plausible metrics along several dimensions to measure the performance of subsystems and the whole system;
- it is a relatively small universe with well-defined boundaries, limited sets of distinct situations and actions, and well-structured state changes over space and time.

When an aircraft flies from one airport to another under instrument flight rules (as civilian and military planes do), it passes through the jurisdiction of a series of ATC centers. These centers track each flight within their sector on radar, and try to keep it on its appointed path and according to a desired time schedule. The control actions must also satisfy a number of constraints. Some are constant, such as the government-prescribed rules for minimum horizontal and vertical separation between planes, and the given physical limitations of the capabilities of the aircraft concerned. Others arise from the situations, such as unfavorable weather conditions and emergency landing priorities. In addition to safe and timely flights, fuel economy, noise pollution over inhabited areas, and proper utilization of air corridors and airport facilities must also be considered.

The *quality measures* of a human or computer ATC operation include the number and the degree of severity of violations of the above constraints. The *competence* of a system is also characterized by the number and the duration of validity of commands — too many and sudden changes in the flight parameters are not popular with pilots, and can lead to accidental or deliberate non-adherence to instructions.

The real-life situation is further aggravated by the uncertainty due imperfect understanding of commands, fuzziness of the aircraft location on radar images, suboptimal commands issued, unexpected environmental changes (weather, unannounced small planes, magnetic storms, etc.). Such uncertainties must be reduced before they lead to unresolvable conflicts.

6.2.2. *Our Approach*

We have adopted the following ideas and assumptions in our work on Distributed Air Traffic Control (DATC) [72, 73]:

- each aircraft has a computer on-board which can communicate with others;
- data gathering/sensing is done by airborne radar equipment;
- communication between aircraft is costly and, at times, noisy;
- each aircraft has uncertain knowledge of the environment;
- there is a natural clustering of activities among the aircraft, both spatially and functionally;
- a general purpose testbed with high-level graphical and I/O capabilities would be a good idea for studying DATC (and other distributed systems);
- the basis of our approach is the *Location-Centered, Cooperative Planning System* (LCCPS) with the following characteristics:
  - the *nodes* are loosely-coupled, with varying degree of autonomy, and belong to decentralized, dynamically formed (possibly overlapping) subgroups;
  - a node can be either a *Nominator*, a *Coordinator* (appointed by a Nominator) or a *Coworker*; it can also have multiple activities in taking different roles — assigning and accepting tasks and sharing computational results;
  - the knowledge base is distributed;
  - the control structure among the subgroups is hierachically organized and is based on Coordinator-Coworker relations.

Nodes are assigned (partitioned) into subnetworks (subgroups) according to the characteristics of the task at hand. The nodes in a subgroup are called 'relevant' to the problem to be solved. The two major factors that affect subgroup formation are the characteristics of the problem (size, complexity, location and decomposability) and the constraints imposed (availability, capability and location of the nodes).

The LCCPS can be viewed as a hierarchical network with one or more subgroups of nodes at each level. This structure acts as a meta-level control which determines the flow of control, data and knowledge. The Coordinator-Coworker relation transforms the individual nodes into an effective and efficient *problem solving team*, and guides the distribution and coordination of the various planning activities.

6.2.3. *On Coordinator Selection*

When a new Coordinator is needed, a Nominator is first selected, which is an available node with the highest ID number — an *implicit agreement* between the nodes relevant to the problem. (The only communication needed with implicit agreements is the exchange of factual information. *Explicit agreements* are made through *negotiations* and confirmation is also needed.) The Nominator, say n', is responsible to nominate, out of all relevant nodes, the best qualified node to be the Coordinator. It is based on the data available in the so-called information packet of the Coordinator candidates:

- the nodes' ID,
- the nodes' Coordinator-Role-Counter,
- the nodes' capability measure.

Node n' sends the Coordinator-Job-Offer to, say, n'' which has the necessary capability and the smallest Coordinator-Role-Counter (as believed by the Nominator). This latter selection criterion is well-justified for "role and load balancing" since Coordinators have to process more messages than Coworkers — they are responsible for distributing tasks and collecting reports. If the *true value* of the Coordinator-Role-Counter of n'' is greater than indicated in the offer (due to delays in passing messages about the status of the network; for example, n'' may have accepted the role of a Coordinator before it processed the message about the offer), n'' does not accept the offer and becomes the new Nominator. This continues until a node finally accepts the nomination and an explicit agreement is made[1].

---

[1]This scheme is more effective in reducing the possibility of communication congestion than that in the Contract Net approach. There, the manager's role is assigned to the node that generates the task (self-appointment). In our case, a Coordinator is appointed on the basis of *true* balanced role assignments.

The Coordinator at the top-level (and recursively at lower levels) tries to achieve group coherence with respect to a goal. It

- decomposes the given problem into smaller subproblems if possible,
- distributes these subproblems to the best-qualified Coworkers,
- collects the reports (subproblem solutions) from the Coworkers and synthesizes the solution.

A Coworker tries to achieve load balancing in the group. Its responsibilities are to

- decide from which Coordinator to request its next task,
- make an explicit agreement about an accepted task,
- decompose the task if it is possible and appropriate to do so (depending on the network status), and
- report its result to the Coordinator.

As stated before, a node may assume the roles of several Nominators, Coordinators and Coworkers at different levels.

6.2.4. *Configuration Diagrams and Tables*

A Coordinator-Coworker hierarchy (CCH) of depth n means that there are at most n levels of the Coordinator-Coworker relationship. We introduce 'configuration diagrams' and the equivalent *configuration tables* to depict the complex roles individual nodes can assume and the relationship between them.

Fig. 6.2.1 shows the configuration diagram of a distributed system with seven nodes and two levels of the CCH. The five nodes relevant to a given problem are marked by thicker circles. A one-head arrow indicates a subordinating relationship between a Coordinator (to which the arrow points) and a Coworker. (Notice that an arrow can also point to its originating node — a Coworker is then its own Coordinator as well.) A two-head arrow indicates a double relationship between the nodes. The node from which the arrow originates is both a lower-level Coordinator and the Coworker of the Coordinator node to which the arrow points. (In case a Coordinator's parent Coordinator is itself, the node in question must be duplicated in the diagram.) Table 6.2.1 is the corresponding configuration table. Our DATC testbed maintains such tables to make sure that the LCCPS simulation is continuously correct.

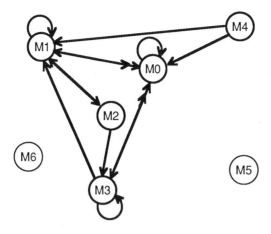

Fig. 6.2.1. — Configuration diagram for a distributed system with seven nodes and a two-level Coordinator-Coworker hierarchy

Table 6.2.1. — An exemplary configuration table corresponding to the above configuration diagram

Node	Role(s)	Levels	Coordinator works for Node	Coworkers belongs to Node
M0	Coworker	0	M0	-
		1	M1, M3	-
	Coordinator	0	-	M0,M1,M2,M3,M4
M1	Coworker	0	M0	-
		1	M1, M3	-
	Coordinator	1	M0	M0,M1,M2,M3,M4
M2	Coworker	0	M0	-
		1	M1, M3	-
	Coordinator	-	-	-
M3	Coworker	0	M0	-
		1	M1, M3	-
	Coordinator	1	M0	M0,M1,M2,M3,M4
M4	Coworker	0	M0	-
		1	M1, M3	-
	Coordinator	-	-	-
M5	no role	-	-	-
M6	no role	-	-	-

Figure 6.2.2 shows the configuration diagram of the same distributed system but with a one-level CCH. Here, the node with all the one-head arrows pointing to it is the leader of the group of relevant nodes and has become a zeroth-level Coordinator. Table 6.2.2 represents the corresponding configuration table.

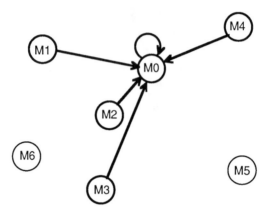

Fig. 6.2.2. — Configuration diagram for a distributed system with seven nodes and a one-level Coordinator-Coworker hierarchy

Table 6.2.2. — An exemplary configuration table corresponding to the above configuration diagram

Node	Role(s)	Levels	Coordinator works for Node	Coworkers belongs to Node
M0	Coworker	0	M0	-
	Coordinator	0	-	M0,M1,M2,M3,M4
M1	Coworker	0	M0	-
	Coordinator	-	-	-
M2	Coworker	0	M0	-
	Coordinator	-	-	-
M3	Coworker	0	M0	-
	Coordinator	-	-	-
M4	Coworker	0	M0	-
	Coordinator	-	-	-
M5	no role	-	-	-
M6	no role	-	-	-

6.2.5. *Connection through Mutual Selection*

In the Contract Net, the *announce-bid-award* form of negotiation is used to find the most appropriate contractor to execute a task. When a potential contractor receives the task specification from the manager, it uses its so-called Knowledge-Source-Centered (KSC) information to decide on its own relevancy to bid for and, possibly, to execute the task. The manager, in turn, uses the so-called Task-Centered (TC) knowledge to decide on which contractor to award with the task. The KSC and TC knowledge are jointly referred to as meta-knowledge.

In our Coordinator-Coworker structure, a similar *announce-request-assign* mode of negotiation is used to solve the connection problem. However, connections are made through *agreements* rather than contracts. It has several advantages to be discussed next.

A potential Coworker receives several *task-announcement* messages, which it then processes to decide whether to submit a request for any of the tasks available and, if so, to which Coordinator. The *task-request* message from a potential Coworker to a Coordinator includes a preference-ranking over all tasks for which the Coworker is competent. The ranking is determined through a relevancy evaluation based on the Coworker's KSC knowledge. Further, it uses its TC knowledge to decide which Coordinator should receive the task request.

The Coordinator uses its KSC knowledge to decide about the order in which the tasks should be handled. It then employs its TC knowledge to assign the task to a specific Coworker, based on its qualifications. The process is repeated until there is no more task to assign or there is no potential Coworker left.

The number of messages necessary is significantly reduced with this arrangement and the process is also very effective. We have also made a small modification in order to decrease the effort needed in processing the long task-available and task-request messages. The task-available messages now contain only a general description of the task. The potential Coworker attaches to the task-request message a brief description of its own capability (such as skill levels). The Coordinator now evaluates the relevancy of the task requester — a small amount of extra work in return for much shorter messages and communication delays.

Load balancing is achieved though the heuristic of *scarce resources*. Since certain tasks can be executed only by a few nodes and certain nodes can execute only a few tasks, a higher priority is given to such more difficult connections. Another important aspect of our approach is that the Coworkers can also *initiate connections*. Namely, a busy Coworker does not process all task-announcement messages right after receiving them. It stores on a *task-sources-list* the IDs of the Coordinators that have sent the messages. (The corresponding ID is eliminated from this list when a Coordinator "resigns" after having assigned a task.) When the Coworker has finished its current task, it will send a task-request message for the most suitable next task obtained from the list. This arrangement reduces the total number of messages necessary (the Coordinator no longer keeps announcing available tasks), decreases the delays in establishing connections and improves the quality of connections.

6.2.6. *Distributed Scratch Pads and the 'Self-Heal' Process*

We call the data structure used to support the above arrangement *Distributed Scratch Pads*. Each Coordinator has a scratch pad associated with each task, on which the decomposed subproblems are posted and onto which the Coworkers' solutions are reported back. The Coordinators synthesize the subproblem solutions at their level and report them back one level higher. These are then filtered back level by level to the top level Coordinator which produces the whole solution.

Distributed Scratch Pads provide a faithful description of the important activities and the current status of the problem solving process. They also help in recovering from individual *processor failure* at a level other than the top level. A Coordinator can discover a suddenly non-functioning Coworker and assign its task to another qualified Coworker. Similarly, a parent Coordinator can detect a subordinated Coordinator's failure, recover the relevant scratch pad information and appoint a new subordinated Coordinator. We call this approach the *self-heal* process. It is not too time-consuming and very reliable when the CCH has several levels and the number of nodes is fairly large.

6.2.7. *Definitions of Incidents, Conflicts, Violations, and Space Measures*

An *incident* is an unwanted spatial or temporal configuration of aircraft in the air space. It can be either a conflict or a violation. A *conflict* is a situation in which two or more aircraft get too close to each other. A *violation*, in contrast, involves only one aircraft that violates some predefined flight parameter. There are two types of violation. A *Type I violation* is shown, for example, in a simulation-based planning system at some future point of time. Such violations must be resolved before they can occur in real life because they can have very severe consequences. An example is the 'Final-Approach-Altitude Violation', which means that an aircraft in the landing phase is flying too high to complete its activity before the end of the landing strip. A *Type II violation* can be resolved before, during or after its discovery. An example is the Unheeded Request Violation, when a pilot has not followed a controller's request for some reason.

The *resolution plan* to resolve incidents must be coordinated since in trying to avoid an incident, others may also be created (or eliminated). *Negotiations* come into play in case of conflicting interests. Further, appropriate updated *plan segments* have to be issued to each of the affected aircraft to modify their current plan if necessary.

A number of important definitions, based on the above concepts, will be needed. The aircraft that may affect or be affected by a resolution of a potential incident are called *relevant aircraft*. Let the *radar range* of aircraft p be denoted as $R(p)$. The space covered by it is called the *Aircraft Surrounding Space* (ASS) and other aircraft within this space constitute the *Surrounding Aircraft* of p. The *Incident Space* (IS) of an incident I is the union of all the Aircraft Surrounding Spaces of the incident's participants,

$$IS(I) = \bigcup_{<\forall p_i \text{ in } I>} ASS(p_i);$$

In resolving incident I, it is important to consider the incidents in the vicinity of I because these may affect its resolution. A reasonable measure for it is the union of all the incident spaces of a group of nearby incidents that also include I as a subset — the *Adjacent Incident Space* (AIS). To obtain a mathematical formalism for it, let us define first the *Adjacent Incident* (AI), I, of aircraft p:

- p is not a participant of I, and
- there is at least one participant of I that is in the Aircraft Surrounding Space of p.

An incident I_i is considered to be an Adjacent Incident of another incident I_j if and only if one of the participants of I_i is in the Incident Space of I_j. Thus all Adjacent Incidents of the participants in an incident I are Adjacent Incidents of I. An aircraft can, therefore, identify its Adjacent Incidents by asking its surrounding aircraft for the list of incidents they are involved in. Moreover, the participants of a potential incident I can find the Adjacent Incidents of I by exchanging their Adjacent Incidents information with each other.

The space covered by incident I and its Adjacent-Incidents is the *Adjacent Incident Space of Incident I*, AIS(I), and is expressed as follows

$$AIS(I) = \bigcup_{<\forall p_i \text{ in } I>} AIS(p_i) \; \cup \; IS(I);$$

where AIS(p) is the *Adjacent Incident Space* of aircraft p

$$AIS(p) = \bigcup_{<\forall \text{Adjacent-Incident } I_i \text{ of } p>} IS(I_i);$$

It should be noted that all aircraft relevant to incident I are in AIS(I). All these concepts will be necessary in discussing the planning process.

6.2.8. *The Kernel Design of the Individual Processors*

We assume that all aircraft are functionally identical and that communication between different aircraft has a constant and limited bandwidth. The messages are error-free and are sent unprompted (without request). In turn, no acknowledgement is given or expected.

Figure 6.2.3 shows the kernel design of an individual airborne processor, a *Distributed Planner*. Information is obtained by an aircraft through *communication* and by *sensing* (radar). Each aircraft has a limited sensory horizon — within the range of its radar. Its knowledge of the world is restricted, uncertain and changing while it and the other aircraft move through the air space. The *Plan Generation Unit* performs a certain "width" (involving a number of aircraft) and "depth" (time span) of planning. The *Look-Ahead Unit* simulates future events, calculates extrapolated trajectories, and consults with the *Incident Detector*. The latter, in relying on the safety rules stored in the *Knowledge Base*, identifies and reports the type, participants,

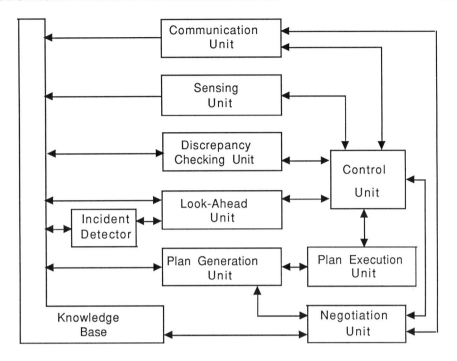

Fig. 6.2.3. — The kernel design of the individual airborne processors

time and location of all incidents. When an incident is detected, the Look-Ahead Unit directs the Plan Generation Unit to modify the current plan. Such modification is also necessary when the updates from the *Sensing Unit* are at variance with the expected world in the Knowledge Base, as discovered by the *Discrepancy Checking Unit*. The *Negotiation Unit* distributes the work load among the affected aircraft and does conflict resolution if necessary. Finally, the *Control Unit* coordinates the above processes.

6.2.9. *The Mechanism to Ensure Navigational Safety*

The planners look ahead in the future, up to a certain *Look Ahead Time* to detect potential incidents and to modify existing plans if necessary. The objective is to have each planner continuously possess an up-to-date knowledge of future situations within the present Look Ahead Time. If there is a potential incident, the planner initiates its resolution. At the end of every short time period, the *perception interval*, it checks for any beyond-tolerance-level discrepancy between the current and the expected worlds. (The latter is obtained by extrapolating from the actual world at the beginning of the last look-ahead process and is stored in the Knowledge Base.)

In case of a larger than tolerated discrepancy, the look-ahead process starts for a full period of Look Ahead Time. Otherwise, only the period of a perception interval is used for the process. The small cost of storing the extrapolated world at the end of every look-ahead process is well-compensated by the several advantages of this approach. These are as follows

- There is no *ad hoc*, fixed value for the Look Ahead Time. (If it is too short, the planning process wastes precious computer time; if it is too long, the planner may not have enough time to derive a satisfactory plan to resolve potential incidents to occur in the *near future*.)
- Redundant calculations are avoided.
- Immediate responses are obtained when they are most likely to be needed.

The above also serves as a design criterion for the on-board radar. The radius of the radar range must be sufficiently large so that the Look-Ahead-Unit can spot potential incidents early enough for their resolution to be guaranteed. Potential incidents are checked every T seconds where

$$T = \min \{30, t_{crit}\}$$

with

$$t_{crit} = \min \{\text{distance } (p_i, p_j)/(\text{max-speed}(p_i) + \text{max-speed}(p_j))\}$$

In other words, the timing is the lesser of the two: thirty seconds or a critical time determined as the period required for the two aircraft involved, p_i and p_j, to hit each other at their maximum speed.

The perception interval is the larger of the two: processing time needed for the look-ahead or the regeneration time of the radar image.

6.2.10. *The Priority-Factor and Its Use*

The question arises: In resolving a potential incident, which aircraft should change its flight parameters and when? We define the *Urgency Factor*, U_i, of aircraft p_i as (-1) times the shortest time period in minutes after which a conflict would occur. (The reason for this form is discussed in connection with the definition of the priority factor.) The *Status*, S_i, of aircraft p_i is defined as the sum of the *Emergency-Level* (critically ill person, hijacking attempt, etc.), E_i, and a measure of the fuel still available, G_i

$$S_i = E_i + G_i$$

Here, E_i and G_i, independently of each other, can assume the values 2, 1 and 0 (normal). We have also devised a measure for the (weighted number of) incidents each aircraft is involved in. It is as follows:

$$W_i = V_i + \sum_{k=1}^{C_i} n_{i,k}$$

Here, V_i is the number of violations p_i is involved in; $n_{i,k}$ is the number of aircraft involved in the k-th conflict of aircraft p_i, $k=1,2,\ldots C_i$; C_i being the total number of conflicts of p_i.

Let the number of aircraft surrounding p_i be N_i. The *Priority Factor* is then with the above defined entities

$$f(p_i) = 8^*U_i + 4^*S_i + 2^*W_i - N_i$$

The weighting factors in the above formula express the proper preference ranking in the negotiation process. The higher the value of the Priority Factor is, the more critical it is to change the flight parameters of the aircraft at hand. Also, a higher value is associated with a higher productivity (effectiveness) in eliminating *all* incidents by changing the course of the aircraft so qualified. In summing up

- the aircraft so identified has the most degrees of freedom;
- this aircraft will need the least number of subsequent messages about later course changes;
- the number of other aircraft affected by the commands for this aircraft is the smallest.

It should also be noted that the values of the terms in the formula for the Priority Factor can be transmitted between the aircraft in question through radar signals in the sensing phase. The whole "computation" is based on simple table look-ups, shifting registers and adding them up. Thus, the negotiation process needed for the preference ranking does not involve the Communication Unit, and the cycle for the request-response can be eliminated. The above formalism leads to an ordered list of the aircraft to be considered in resolving a potential incident.

It is possible for an aircraft to detect several incidents during the look-ahead process. For the purpose of ordering such incidents to be resolved, we have defined the *Incident Criticality Factor* as the sum of the Priority-Factors of all aircraft involved in the given incident,

$$F(I_j) = \sum_{<\forall p_j \text{ in } I_j>} f(p_j)$$

The most critical potential incident detected by a particular aircraft is called the *Individual Most Critical Potential Incident*. For the benefit of reducing both the number and the severity of possible *subsequent* incidents in the most productive way, we have also defined the *Group Most Critical Potential Incident*, which has to be resolved first.

The overall advantages of the above approach are:

- we can avoid commands that would potentially lead to conflicts;
- it is easier to coordinate the resolution of co-existing incidents between groups of relevant aircraft;
- the system can allocate better the available processing resources to the resolution of potential incidents.

6.2.11. *The Incremental Shallow Planning*

The *planning paradigm* we have adopted is simulation-based *Incremental Shallow Planning* (ISP), a multi-path and multi-stage search process. At each pass, a 'satisficing' (as opposed to optimum) plan is derived for a given world situation. Further passes are performed to modify the existing plan when new information is obtained about the environment. The number of stages for a pass is determined either by the time allowed for a pass or for the subgoal to be achieved.

Shallow planning is characterized as a search of a limited-depth tree. The search can be blind or heuristically guided. The root node is selected from among the nodes generated in the previous stage. The depth of the tree (that is, the depth of shallow planning) is constrained by the time allocated to planning for that stage. The outcome of the previous shallow planning guides the search for the total solution or the partial solutions at each stage. The entire planning process terminates when the final goal is accomplished.

The technique is especially effective for a group of distributed agents planning cooperatively when

- the search space is large (many alternative actions are feasible at each juncture point);
- the planner's knowledge of the world is incomplete and uncertain;
- the planner has limited control over the environment.

Each of the various possible incidents is associated with appropriate *directives* in the knowledge base. The planner has to identify

- which directive is to be used and with what parametric values,
- the aircraft to use the directive, and
- the time of using the directive.

A directive with tentative parametric values is evaluated in the simulation phase of planning. Different individual agents can be "in charge" of the independent and simultaneous simulation activities along the different branches of the 'DATC-Search-Tree'. (This idea gives rise to the notion of the Distributed Scratch Pads and fits naturally the Coordinator-Coworker structure described earlier.) Once the directive is found satisfactory, it becomes a *Command* — that is, an *Instantiated Incident Resolution Directive*.

6.2.12. *The Coordinator-Coworker Structure as the Organizational Scheme*

Let us see how a node with multiple Coordinator roles assigns the tasks of evaluating directives to potential Coworkers. In processing *task-request* messages, it first considers all zeroth-level incidents with unprocessed tasks. If there is at least one, it

- selects the *most* critical incident (by using its KSC knowledge),
- checks (by using its TC knowledge) whether the task requesters are relevant to the incident at hand (this is necessary because of the node's multiple Coordinator roles),
- chooses the most appropriate Coworker and sends a *task-assignment* message to it.

This process continues until either there is no more zeroth-level incident with unprocessed tasks or all task requesters have gotten their assignments.

If there are still some task requesters without assigned tasks, the Coordinator starts considering first-level incidents with unprocessed tasks. If there are none, the Coordinator sends out *no-task-available* messages to the remaining task requesters. Otherwise, the Coordinator

- selects the *least* critical first-level incident,
- checks whether the task requesters are relevant to the incident at hand,
- chooses the most appropriate Coworker and sends a *task-assignment* message to it.

Selecting the least critical first-level incident means that the Coordinator is aiming at the most promising direction in the search for the best resolution of the zeroth-level incident at hand. (The reason is that the quality of the resolution of the zeroth-level incident is measured primarily by the criticality of the immediate first-level incidents it causes, if any.)

The above process continues until either there is no more first-level incident with unprocessed tasks or all task requesters have gotten their assignments. Finally, a *no-task-available* message is sent to every remaining task requester without an assignment. The Coordinator has used its KSC and TC knowledge in assigning the most appropriate task to the relevant available Coworkers.

6.2.13. *The Process of Resolving Incidents*

A Coordinator is first selected to direct the resolution of a zeroth-level incident. It creates a zeroth-level scratch pad and posts on it the branches of the search tree to be developed further. The need for simulation is then broadcast to the relevant aircraft. The task are assigned, the respective commands are computed at that level, and the new branches of the search tree become available for further development. First-level Coordinators with level-one scratch pads are selected, and the same process of broadcasting the need for simulation, assigning tasks and computing commands is performed. The images of the Extrapolated World, resulting from all possible commands, are then evaluated. The best ones are sent back level by level to the top-level scratch pad. Finally, the zeroth-level Coordinator makes its selection of the commands to be used along the solution path and broadcasts its decision.

Next, the problem of resolving co-existing potential incidents needs to be addressed. If an aircraft's Individual-Most-Critical-Potential-Incident (IMCPI) is the Group-Most-Critical-Incident (GMCPI) of its group of relevant aircraft (as defined in Section 6.2.10) then that aircraft becomes the Coordinator for resolving that incident. Otherwise, it is a Coworker, a member of the problem solving team of all relevant aircraft. First, the resolution of GMCPI is accomplished, which may also resolve other IMCPIs at the same time. The Coordinator sends the resulting *new-plan-segment* message to all relevant aircraft. These latter start their perception phase immediately to detect their IMCPI. The whole group planning process is repeated iteratively until there is no potential incident left.

6.2.14. *The Three Organizational Structures To Be Compared*

We have carried out a series of experiments to measure the respective levels of control performance in three organizational structures:

- the Local, Centralized Architecture (LCA),
- the Location-Centered, Cooperative Planning System with One-Level Coordinator-Coworker Hierarchy (LCCPS-1), and
- the Location-Centered, Cooperative Planning System with Two-Level Coordinator-Coworker Hierarchy (LCCPS-2).

It should be noted that LCA can also be considered to have a zero-level Coordinator-Coworker Hierarchy. The comparisons have been made in terms of communication overhead and processing time needed for planning — all at different levels of traffic density and problem size.

The Incremental Shallow Planning generates the DATC-Search-Tree of up to two levels of branches. The development of a branch is called a *task*, and the task level corresponds to the branch level. Let N_1 and N_2 be the number of first and second level tasks, respectively. The processing time needed to generate the DATC-Search-Tree with the LCA is

$$\sum_{i=1}^{N_1} t(T_i) + \sum_{j=1}^{N_2} t(T_{i,j})$$

where $t(T_i)$ is the time needed to complete task T_i, and $T_{i,j}$ is the j-th level-2 subtask of the level-1 task T_i.

Let us now assume that there is an unlimited number of processors available. The processing times needed to generate the DATC-Search-Tree for LCCPS-1 and LCCPS-2 can be calculated as

$$\underset{<i=1,...N_1>}{\text{Max}} \left\{ t(T_i) + \sum_{j=1}^{N_2} t(T_{i,j}) \right\}$$

and

$$\underset{<i=1,...N_1>}{\text{Max}} \left\{ t(T_i) + \underset{<j=1,...n_i>}{\text{Max}} \{t(T_{i,j})\} \right\}$$

respectively. Here, n_i is the number of level-2 subtasks of the level-1 task T_i.

As can be seen from the above formulae, relatively more parallel activities become possible in the LCCPS-2 when an unlimited number of processors are available. *For the time being*, let us accept the following simplifying assumptions:

- the execution time of every task is identically equal to t;
- the number of level-2 subtasks of each level-1 task is identically equal to n (that is, $N_2 = n*N_1$);
- the time needed for selecting a Coordinator is negligible;
- the message transmission time for assigning tasks and reporting results is negligible;
- every message is responded to immediately after it has reached its destination.

In such an idealistic environment, the processing times needed for planning in the three organizational structures and the speed-ups obtained in using LCCPS-1 and LCCPS-2 are shown in Table 6.2.3. However, as we have found out, none of the above assumptions hold exactly and the exact duration of the respective planning processes cannot be formulated analytically. Therefore, we have designed and implemented a general-purpose testbed for studies on distributed planning and problem solving. With reference to the experiments to be described below, we refer to it as the DATC testbed. The results of the experiments performed in the testbed have enabled us to gain a better quantitative understanding of the impact of the contributing factors on the overall effectiveness of the three architectures.

Table 6.2.3. — Processing times needed for planning and the speed-ups obtained with the three organizational structures in the idealistic case when the simplifying assumptions described in the text hold

Architecture	Processing Time Needed	Speed-Ups Obtained
LCA	$(N_1 + N_2)*t$	-
LCCPS-1	$((N_1/N_2)+1*t$	$N_1 - 1$
LCCPS-2	$2*t$	$((N_1+N_2)/2) - 1$

6.2.15. *The Distributed Air Traffic Control Testbed*

The objective of the DATC testbed is to provide a powerful and effective simulation environment for empirical studies on different DATC architectures.

We consider multiple aircraft behavior at two levels. The lower level is navigation oriented. The higher level is concerned with the elimination of incidents and is, therefore, strongly influenced by the imposed organizational structure. For example, when an aircraft is to descend from a certain altitude, it follows a sequence of actions — a lower level behavior. A higher level behavior can be witnessed when an aircraft detects a potential incident. It is characterized by its interaction with other affected aircraft and the consequences of such interaction. The higher level behavior also controls in our case the distributed planning processes — for example, the planning responsibilities delegated to the participating aircraft.

We assume that each aircraft in the testbed uses the same *Distributed Planner* to control its flight and to ensure its navigational safety. Further, each aircraft is capable of performing sensing, look-ahead, potential incident detection, plan generation and execution, communication and, if appropriate, negotiation. An aircraft can be one of three types: 'superjet' (short for supersonic jet), 'jet' and 'propeller'. Each category has different performance characteristics, such as climb speed, maximum speed, maximum altitude, etc.

Every aircraft knows its own *flight plan* and that of the other participating aircraft. Due to the interaction between aircraft and unforeseen circumstances (e.g., adverse weather conditions), flight plans are deliberately not completely adhered to. When that happens, the aircraft follow commands generated by various planning processes. These commands are not known to other aircraft unless they are requested through communication.

Each aircraft undergoes different phases of behavior during its flight, such as take-off, ascending, cruising, descending, holding, approach-preparation, approaching and landing. The aircraft have a temporary goal in each phase (e.g., to ascend to a specified altitude, to cruise at a given speed, etc.) and some expected behavior.

We distinguish between three different worlds in the testbed. The *Real World* (RW) reflects the situation in the global airspace and monitors the performance of an organizational structure. The rich associated knowledge base contains all aircraft's flight plans and current flight parameters, and the commands that can be issued. No incidents occur in this world — they must be detected and resolved by the aircraft involved in advance.

Every aircraft maintains the image of a *Simulated World* (SW) which reflects its surrounding environment on its 'radar scope'. It is obtained by extracting the relevant information from the RW directly.

Finally, every aircraft creates and uses various *Extrapolated Worlds* (EWs). EWs are used to detect discrepancies between expected and actual situations at appropriate time points, to look ahead to estimated future states, and to detect potential incidents.

The major components of the testbed are:

- the User Interaction Module,
- the Planner Module,
- the Navigator Module,
- the Database Module,
- the Graphics Display Module,
- the Incident Detection Module,
- the Message Processing Module,
- the Debugging Module.

The testbed can be run

- in the graphic (three-dimensional, perspective view of one of the worlds) or non-graphic observation mode (declarative description of a world);
- in the batch or interactive running mode;
- in trace modes of three different levels of detail;

- in three different initiation modes: 'fresh start', 'fast-return' to an earlier time, or 'checkpoint' to last finished status;
- in the presentation mode combining RW, SW, and EW.

There are numerous difficulties in simulating on a uniprocessor distributed, asynchronous planning processes, concurrent events and activities. These difficulties have been overcome by the message passing capabilities of each aircraft and by the proper treatment of time. The latter involves

- the recording of the duration of each activity, including message sending and processing, in terms of actual (real) time;
- the maintaining of the correct RW time from the perspective of each individual aircraft — after each updating of the RW, and receiving and processing a message;
- the integration of the above two functions in the simulation of multiple aircraft activities.

There are in the testbed several general-purpose facilities provided for these objectives and for the updating of the RW.

6.2.16. *The Results of Empirical Investigations*

We have prepared 18 scenarios which differ from each other either in the number of participating aircraft (three different traffic densities) or in the number of branches on the DATC-Search-Tree (two problem sizes) or in the architecture (three organizational structures). The notation DdSsAa has the following meaning: D, S and A indicate traffic density, problem size and architecture, respectively. The lower case letters d, s and a stand for numbers, as shown in Table 6.2.4.

It should be noted that the chosen configurations in the scenarios were carefully considered so that meaningful results should be obtained on the available hardware (a large VAX 11-780) within reasonable running times.

Next, we investigate the timing requirements for messages. We have assumed a baud rate (the maximum number of bits that can be transmitted per second) of 9,600. If ten bits are needed to represent a character, the transmission time for message m is

$$(message\text{-}length(m)*10)/9600$$

Table 6.2.4. — The meaning and possible values of the parameters d, s and a according to the scenario reference convention

Symbols	Values	Meaning
d	2	two aircraft
	3	three aircraft
	4	four aircraft
s	1	two level-1 and four level-2 tasks
	2	three level-1 and six level-2 tasks
a	0	LCA
	1	LCCPS-1
	2	LCCPS-2

We have defined the message-delay-in-response as the difference between the message-response-time and the message-receipt-time.

The statistics collected include

- the number of external messages generated and processed,
- the average transmission time for a single external message,
- the average delay-in-response of a single message,
- the average propagation time of a single message,
- the duration of the Incident Resolution Process,
- the duration of the Incremental Shallow Planning Process,
- the duration of the Relevant Aircraft Identification Process,
- the average degree of processor utilization during the Incremental Shallow Planning Process,
- the average speed-ups possible with the architectures LCCPS-1 and LCCPS-2 for both problem sizes specified.

We have found some somewhat anomalous results concerning the comparisons between the architectures LCCPS-1 and LCCPS-2 in the first runs of experiments. The management of the available processors was suboptimal, planning processes were often held up due to a delay-in-response to messages. This happened under two conditions. First, when a Coordinator is executing a task, all task-requesting messages must be temporarily suspended. This places the task-requesters (the available processors) in a waiting state. Second, the confirmation process of a Coordinator will come to a halt when the nominee is executing a task. This also slows down the planning process considerably.

In order to minimize the probability of occurrence of the first type of situation, we have added more intelligence to the processors in terms of various heuristics to be followed by the Coordinators. (For example, "do not request a task unless all other aircraft have been assigned a task".) With reference to the second reason for the anomalous results, one could either let the processors be interruptable or eliminate the nomination process for resolving level-1 incidents. We have followed the second avenue. When a Coworker detects a new level-1 incident, it becomes a self-appointed Coordinator for that incident within the LCCPS-2 architecture.

We have identified two further minor causes for some inefficiency in planning. First, when a Coworker wants to request a task from a group of Coordinators at a certain level it works for, it uses a 'scheme-of-equal-priority' in selecting a Coordinator from the task-source list. If the Coworker, itself, is on its own task-source list (that is, it has the roles of a Coordinator and a Coworker at the same time), it may end up requesting a task from another Coordinator. This would result in the generation and processing of external messages with task-requests and task-assignments, which would in turn decrease the average processor utilization and slow down the planning process. Our solution is that such a node with double role must request a task first from itself whenever possible — internal messages take no transmission time.

The second minor reason for inefficiency is that even when a given Coordinator knows that it has no more tasks to be executed, it will not automatically notify the other nodes about this situation. This may create many cycles of task-request and no-task-available messages, increasing the idle time of task-requesters. We have, therefore, made the Coordinators notify the other aircraft *as soon as* they are assigning the *last* unassigned task — sending a 'no-other-task-available' message.

It is of interest to give a qualitative account of the results obtained with the above arrangements. We have found that

• the architectures for distributed planning, LCCPS-1 and LCCPS-2 are superior in efficiency to the Local, Centralized (LCA) architecture;

- a careful message passing management reduces the average delay-in-response, and the relevant aircraft become a more effective and efficient problem solving team;
- the total number of messages generated and processed is less with LCCPS-1 than with LCCPS-2;
- the heuristics introduced have reduced the average delay-in-response for single messages, drastically decreased the time required for the Incremental Shallow Planning, and resulted in even more significant speed-ups of the distributed architectures;
- a careful sequencing of task-request messages, as produced by the heuristics, has further increased the processor utilization rate — a result causally related to the drop in the average delay-in-response value;
- when there are not enough processors to take level-2 tasks, LCCPS-2 turns out to be inferior to LCCPS-1 for larger problem sizes[1].

6.2.17. *Conclusions*

The Location-Centered, Cooperative Planning System has proved to be a feasible, effective and efficient organizational structure for DATC. The Coordinator-Coworker paradigm makes sure that the available processing resources are (almost) optimally utilized. Coordinators are nominated according to their qualifications and with reference to their current load as well. This approach reduces the possibility of communication congestion in general. The management of message passing is of great importance. Connections are made through a refined mutual selection process that decreases the number of messages needed, yet enhances the flexibility of the whole system. Connections can be initiated by both Coordinators and Coworkers — an important aspect of distributed control. The Distributed Scratch Pad is a useful tool for partitioning the knowledge base on a 'need-to-know' basis.

The architecture to be chosen for a real-life implementation is a function of the computing power and processing time available. The communication

---

[1]Which is the better architecture actually depends on the difference between the number of relevant aircraft and the number of first level branches in the DATC-Search-Tree. Since each first level branch is created to solve a partitioned subproblem, partitioning of a given problem also plays an important role in efficient group planning. Unfortunately, it was not possible to collect simulation data which describe the behavior of a higher number of participating aircraft to investigate this issue further. (Some of the more demanding scenarios required about 9 MByte core memory and used over 1,400 LISP functions.)

overhead in distributed planning is well-justified. Beyond the issues of flexibility, responsiveness and effectiveness, reliability is also significantly higher within the distributed problem solving environment (cf., 'self-healing'). The DATC testbed has proven to be a valuable tool for studying a variety of phenomena as it displays not only the lower level navigational aspects of the air space but also lets the user monitor the control activity of the distributed planning process.

Finally, a few words should be said about the *costs* involved in the proposed system. With the LCA approach, a single, ground-based large processor would control the air traffic. The power needed for resolving a potential incident, W', is related to the power of each individual air-borne processor, w, working in the LCCPS-2 organizational structure as

$$W' = w*((N_1+N_2)/2)-1)$$

where N_1 and N_2 are the number of level-1 and level-2 tasks, respectively. If there are at most r potential incidents to be resolved simultaneously, the ground-based processor must have $W=r*W'$ computing power.

Grosch's law (not universally accepted these days) says that computing power is proportional to the square of the cost of the computer. It means that, for example, if there are two computers, one twice as expensive as the other, the more expensive one will have four times as much power as the less expensive one. If an airborne processor's price is c, the ground-based system will cost

$$C = c^*(N_1 + N_2)\sqrt{(r^* w((N_1 + N_2)/2 - 1)}$$

For example, if N_1=5, N_2=29 and r=16, the cost of the ground-based system is 34*c versus 16*c for the distributed system. (Note that r can be much larger, too.)

6.2.18. *Future Research Directions*

The following ideas seem to be worth pursuing:

- adding a priority level to each message would enable more urgent messages to be processed earlier;

- adding more intelligence to the Message Processing Module so that those messages that cause a higher productivity of the whole system are responded to earlier;
- enabling some of the planning processes to be interrupted so that certain messages can be responded to faster, which would improve processor utilization;
- studying the effect of noise in message transmission;
- studying the effects of an arrangement in which the number of levels employed in the Coordinator-Coworker hierarchy depends on the situation;
- implementing a multi-processor version of the DATC testbed for simulating "real" distributed planning and problem solving.

6.3. A Distributed System for Manufacturing Control

6.3.1. *Introduction*

We have wanted to establish a system that can control the operation of, for example, a nation-wide distributed manufacturing organization [67, 68, 69]. As will be seen later, the paradigm in question is closely related to numerous other tasks, such as the control of an electrical power network, a collaborative computer network, a nation-wide car rental company and of other organizations. We will, however, use a terminology that is applicable to the issues of manufacturing.

The characteristic features of this environment are as follows:

- The constituent plants possess both a *spatial specialization* (in terms of their geographical location and connectedness) and *functional specialization* (in terms of equipment, capabilities, skills and cost functions).
- The plants are connected by a *transportation network and a computer network* of limited capacity and varying reliability.
- *Cooperation between the plants* is crucial both in the plan generation and plan execution phases.
- The *information* on which planning is based is *volatile* and *incomplete* due to a lack of global knowledge and dynamically changing conditions (new orders are coming in, new plants are established, manufacturing equipment and computing and transportation facilities may break down or are being maintained, cost functions change, etc.).

- There is a *critical dependency* and possible *conflicts*, *incompatibility* and *contests* between subsystem controllers.
- The *communication* is asynchronous between the plants. The messages can be broadcasted at large, aimed at selected groups of plants or at an individual plant.
- The manufacturing operation of a given product defines a *hierarchical network* which corresponds to (is homomorphous with) the problem solving network needed by the planning process. This network may have to be changed at any point of time in responding to new conditions, foreseen conflicts and emergencies.
- The *availability* and *cost functions of resources* (raw materials, supplied components, equipment, transportation and computational facilities) may change intermittently or regularly. Idle plants and setting-up operations also cost money.

There are two possible objectives. The *manufacturing combine* of all plants may have to produce a given number of final products

(1) either at minimum overall cost within a given period of time,
(2) or at a given cost figure within a minimum period of time.

Either of these criteria requires an optimum allocation schedule of the processing/assembly operations to individual plants over space and time, while satisfying a set of current and estimated constraints.

The above described, fairly general-purpose task is a rich domain for studying the problems of connection, timing, network perception, load balancing, message reduction and others. We have created a user-friendly programming environment — partly based on the testbed developed for the investigation of distributed air traffic control — which has the following capabilities:

- it enables the user to specify the problem (characteristics of the products, plants and transportation) in a fast and error-free manner;
- any part of the information given can be modified on-line;
- the user can monitor and evaluate plan generation and execution in either a visual or a descriptive mode;
- the built-in learning facility can rely on automatic and user-guided processes to improve system performance.

We have adopted again the idea of a decentralized, loosely-coupled collection of problem solvers, each located at a distinct network node. The flow of control and information is now defined by the manufacturing process of given products. The *hierarchical network structure* for a hypothetical product is shown in Fig. 6.3.1. It enhances multi-agent planning. Nodes at a higher level have more global knowledge of the problem solving environment and make more general plans than those at a lower level. Nodes are said to be *semi-autonomous*, which means that the mode of decision making is dependent on the *network status* — 'external decision makers' are used when the need for communication and computation is relatively small, whereas 'internal decision makers' are at work when the urgency of subtasks is not an issue and there is no danger of duplicating certain processes.

Planning can be considered as the (constrained) process of selecting a subset of the *Cartesian product* of three performance-related sets,

WHAT x WHERE x WHEN

• WHAT denotes the set of manufacturing/assembly responsibilities of all components, every element of which must appear in the subset;

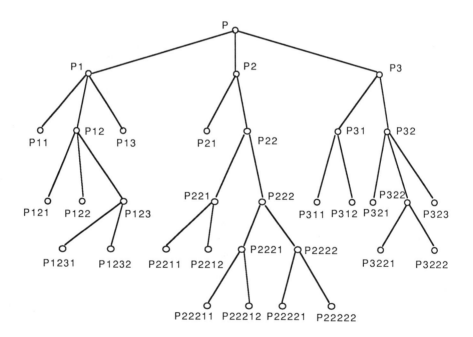

Fig. 6.3.1. — The hierarchical structure of product P

- WHERE refers to the set of plant locations, some elements of which may not be selected and some may be selected several times;
- WHEN references a set of times.

The constraints refer to the optimization criterion used. There can be a number of additional difficulties, such as

- some components are common to several products;
- some components can be processed by only a limited number of plants;
- some plants have only a very limited set of skills;
- some components may be needed in large number and have to be produced by several plants;
- beginning and/or finishing time points of processing and assembly may be subject to various restrictions and constraints on an absolute and/or relative time scale;
- storage space and transportation may be limited in volume and cost, idle plants and setting-up operations cost money, etc.

6.3.2. *The General Paradigm*

Scheduling a single factory has been studied extensively (see, e.g., [24, 27, 84, 105, 203]). The domain in our case contains a group of dissimilar plants distributed geographically. Each plant has a computer which is connected to a communication network. Messages can be sent from a given plant to a particular plant, or to a set of plants selected according to some criteria, or broadcasted at large to all plants.

The *plant* X_i (i=1,2,...,k) is characterized by a *descriptor D*, specifying its skills, equipment, current inventory, direct transportation to other plants, and available storage space:

$$D(X_i): [S_1(X_i), S_2(X_i),..., S_m(X_i); E_1(X_i), E_2(X_i),..., E_n(X_i);$$

$$I_1(X_i), I_2(X_i),..., I_o(X_i); T_1(X_i), T_2(X_i),..., T_p(X_i); V(X_i)]$$

Here $S_s(X_i)$, $E_t(X_i)$ and $I_u(X_i)$ denote the s-th skill (s=1,2,...,m), t-th equipment (t=1,2,...,n) and u-th component or raw material (u=1,2,...,o) the plant X_i has, respectively. $T_v(X_i)$ (v=1,2,...,p) stands for the direct transportation connection between plants X_i and X_v (v=1,2,...,k; i≠v). $V(X_i)$ is the storage space available at plant X_i.

Each *skill* has certain *attributes*:

$$S_s(X_i): \quad [N_s(X_i), \ C_s(X_i), \ O_i(X_i,t)]$$

Here $N_s(X_i)$ represents the number of time units needed for executing skill $S_s(X_i)$; $C_s(X_i)$ is the cost of doing so per unit time; and $O_s(X_i,t)$ is the availability status of the skill at time t.

Each *equipment* $E_t(X_i)$ is identified by its name and has the attribute 'availability status' with two values: the total number of units and the currently available number of units.

Each *raw material* or *subcomponent* $I_u(X_i)$ has four attributes: the amount available, unit cost, and unit weight or volume (depending on how it is packaged, transported and used).

Each *transportation connection* $T_v(X_i)$ is characterized by five attributes: the name of the other plant X_i is connected with, the distance between the two plants, unit transportation cost per batch, batch size and batch transportation time.

The *final product P_j* has components and subcomponents over several levels. It is described as

$$P_j: \quad [P_{j,1}, \ P_{j,2},...,P_{j,r}; \ M_{j,1}, \ M_{j,2},...,M_{j,r};$$
$$S_1(P_j), \ S_2(P_j),..., \ S_g(P_j); \ E_1(P_j), \ E_2(P_j),..., \ E_h(P_j)]$$

In this notation, $P_{j,c}$ is the c-th first-level component of the final product P_j ($c=1, 2, \ ... \ r$); $M_{j,d}$ is the amount of $P_{j,d}$ needed for one unit of P_j ($d=1, \ 2,...,$ r); $S_e(P_j)$ and $E_f(P_j)$ are the e-th skill and f-th equipment needed for the processing/assembly of its components.

Each *component* may have *subcomponents* and is then described as:

$$P_{j,c}: \quad [P_{j,c,1}, \ P_{j,c,2},..., \ P_{j,c,w}; \ M_{j,c,1}, \ M_{j,c,2},...,M_{j,r};$$
$$S_1(P_{j,c}), \ S_2(P_{j,c}),..., \ S_g(P_{j,c}); \ E_1(P_{j,c}), \ E_2(P_{j,c}),..., \ E_h(P_{j,c})]$$

Multiple final products require a dummy task acting as the root of the hierarchy tree, the parent node of each final product. An exemplary case is shown in Fig. 6.3.2.

Finally, the *cost* of producing P_j in plant X_i is given as

$$V(X_i, P_j) = C(X_i, P_j) + U(X_i, P_j).$$

Here $C(X_i, P_j)$ is the total cost of producing and transporting to X_i all components one level below P_j; $U(X_i, P_j)$ represents the cost of processing/assembly of P_j at plant X_i. (The cost of planning is assumed to be negligible.) This expression is recursively true of every component at each level; however, only the purchase and transportation costs of raw materials appear in the formula.

6.3.3. *The Approach*

The *negotiation process* enables a node to reveal as much as necessary of its status information to selected others. *Network perception* keeps a node's model of the network up-to-date on the basis of messages sent and received. It covers the status of the network and of the problem solving process.

Some of the *knowledge* is *static* (e.g., the location of a given plant), some *changes infrequently* (e.g., the number and type of machines at a given plant), some *changes dynamically* (e.g., the set of qualified receivers of a given message).

For the planning process, there are two types of strategies possible:

- The *top-down strategy* starts at the root, generates an abstract overall plan and refines it level by level. Its disadvantage is that component costs and timing are unknown within the abstract plan below the level being considered.
- The *bottom-up strategy* starts with raw materials. Its disadvantage is that there are resource allocation conflicts and possibly only locally optimum

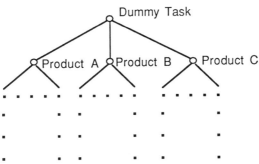

Fig. 6.3.2. — The case of several final products

resource allocations, and the load distribution can be unbalanced. Backtracking is often needed to rectify these problems.

We have selected the latter approach. To reduce resource allocation conflicts, certain heuristics were added to the decision criteria in planning. So-called *reservation rules* mean that the capabilities for a task are reserved if less than a predefined number of bidders occur. Further, we have allowed 'over-bidding' by a node for the sake of better resource utilization. The rule is also kept that a node does not bid for other tasks when it is the single bidder for a particular task *(self-reservation)*. *Negotiation* is kept as the last resort since it may become costly in time needed for sending and interpreting messages.

We have adopted an architecture based on dynamically generated *Contractor-Workers interaction*. The nodes are partitioned into *need-defined groups*. The significant advantage is that negotiation and communication are constrained to smaller regions of the network.

The *Coordinator* (specified by the user for a particular product) receives the problem from an "ordering agent" (usually the user). A top-level *Contractor* is then appointed by the Coordinator. Contractors of lower levels are selected by the Contractor of one level higher. A particular node can have multiple roles over the whole network but, at any single level, it may have only one Contractor role in order to reduce communication problems.

The *problem decomposition* is usually on the basis of AND-OR trees, as shown in Fig. 6.3.3. The system design is shown in Fig. 6.3.4.

Network perception is useful for acquiring the specific knowledge of what has been done, what is being done and what else needs to be done. The *operational cycle* of a node is based on the 'perceive-plan-act' loop mentioned before. Four heuristic rules assist it in selecting the next task:

Rule 1: Tasks located closest to the bottom of the task hierarchy should be selected first since longer paths are more likely to develop into bottlenecks resulting in longer overall response time.

Rule 2: Out of several ready tasks at the same level, the one with finished sibling tasks should be selected. Once this is accomplished, the parent tasks can be considered.

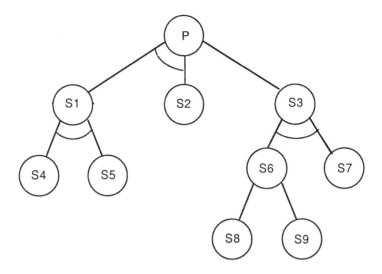

Fig. 6.3.3. — Problem decomposition with an AND-OR tree

Rule 3: Out of several ready tasks at the same level, the one whose parent task is not in the current job queue should be selected in order to balance completed levels as much as possible.

Rule 4: Out of several ready tasks at the same level, the one with the shortest time requirement should be selected.

The individual components have the following responsibilities:

- The *Network Planner*, being in charge of node cooperation, sends and receives messages via the *Communication Unit*. It also decides which node should bid for a given task.
- The *Local Planner* generates the detailed bids using its local status information. The *Meta-Level Planner* refines the high-level, long-term plans into low-level, short-term ones in considering the current network status (e.g., the needs of other nodes).
- The *Perception Unit* comprises the *Message Record Unit*, the *Perception Processing Unit* and the *Perception Model*. The first files the messages sent and received. The second identifies those messages that can cause a change in the Perception Model. It updates the Perception Model only when the Meta-Level Planner directs it to do so.

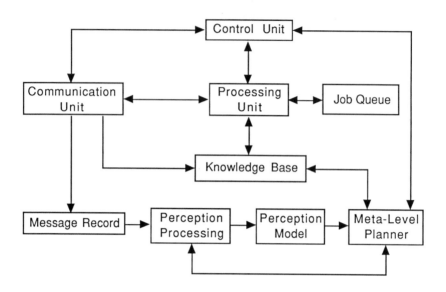

Fig. 6.3.4. — The system design of distributed manufacturing control

In addition, each node has a *Graphics Display Unit*, a *Knowledge Base Maintaining Unit* and a *Plan Execution Monitoring Unit*. The user can operate the system in any one of the following four modes:

- In the *Creating Mode*, the user can specify the manufacturing and transportation environment and the products to be produced (equipment, skills and inventory of plants, transportation routes between plants, costs, product hierarchy, etc.).
- The *Modifying Mode* enables the user to change any of the above, including those items that are automatically identified by the system as affected by a modifying action.
- The user may wish to see on a graphics display and print out any of the above in the *View Mode*.
- The *Operational Mode* receives information from the user about the name and the amount of the products to be produced, desired starting and finishing times, and the manufacturing criterion (time or cost priority).

The correctness and the feasibility of these are then checked by the system.

The manufacturing plan generated can be displayed in several different modes, such as a

- *general plan* (shown are timed tasks assigned to plants),
- *plant-oriented plan* (shown are timed tasks assigned to equipment and skills used, and the suppliers used),
- *component-oriented plan* (shown are plants, equipment, skills used for given components within shown time periods).

We are currently studying an issue of general concern — namely, how to minimize the *extent of replanning* necessary when environmental conditions change.

6.4. A System for Distributed and Moving Resources and Tasks

6.4.1. *Introduction*

We discuss in this section another paradigm of fairly general applicability. It will first be described in general terms but the terminology subsequently used will be with reference to an important application.

Let us consider a *task environment* with the following basic characteristics:

- There is a *hierarchy of decision making entities* — the higher the level of such an entity, the more global and abstract its knowledge is about the current state of the environment.
- Each of the decision making entities at the lowest level of the hierarchy (to be called 'Stations') has a *set of resources* with given characteristics and an area of *basic jurisdiction* for task accomplishment.
- Scheduled and unscheduled *tasks* become known to the Station in whose jurisdiction they occur. These tasks have to be accomplished with varying amounts and types of resources, and within given, task-dependent time periods.
- The resources needed for a given task may not be available at a given Station at the right time — they are either currently assigned to other tasks, are under repair or just not owned in sufficient number by the Station in question. In such cases, resources assigned to lower priority

tasks may have to be *re-assigned* to higher priority ones or resources owned by other Stations must be *turned over* to the jurisdiction of the Station responsible for the task.

- Tasks are usually *non-stationary* and they may also move across the boundaries of Station jurisdiction. The resources assigned to such a task at a given time need not stay with the task but can be replaced by others whose location or other characteristics are better suited to the requirements.
- In turn, as tasks become accomplished, the resources assigned to them become available.
- The properties of the environment change: new Stations may be established; boundaries of Station jurisdiction may change; additional and/or new types of resources can be acquired; new types of tasks may occur; the 'resource management style' of the decision making entities can be modified; physical, legal, fiscal, organizational and other constraints are subject to change.

The following *objectives* must be considered in the environment described:

- The resources must be *allocated* in an optimum manner, over space and time, to scheduled (regular) and unscheduled (unexpected and emergency) tasks. This means that each task is completed by the most appropriate *resource-mix* (at a minimum overall cost) and within a satisfactory time period. Further, the load on the resources must be well-balanced, human resources are not to be overutilized, and repair requirements must also be satisfied.
- Relevant *past events* (tasks, plans, solutions, performance results) are recorded and made use of for a *learning program* aimed at improving performance as more experience is obtained.
- Currently available resources are distributed among the Stations on the basis of the *history of task distributions*.
- The *acquisition of new resources* is done in an optimum manner. In other words, the best resource-mix has to be purchased either within a given budget or for a given set of predicted task-mix scenarios.
- The *problem solving environment* assumes a close man-machine interaction, and provides for the user high-level input and monitoring facilities.
- All *system recommendations* for action must be self-explicatory, and are accompanied by a quantitative and qualitative analysis of the trade-off between cost and accomplishment.

- The human operator can *accept* any of the system recommendations or *over-ride* all with a "manual" solution.
- The system must be able to *support both day-to-day and longer-term operations* (e.g., maintenance of resources, accomplishment of scheduled tasks and responding to emergency tasks versus locating Stations strategically, acquiring new resources and allocating them to individual Stations).
- In case of a partial failure with some communication or computation units, the system must not crash but undergo a *graceful degradation*; that is, it must still provide satisfactory solutions but possibly of lower quality and at a slower pace.
- It should be easy to *expand the environment* and to provide support for new Stations, resources, conditions and constraints.

6.4.2. *A Domain of Application*

Next we discuss an important application of the above ideas. It involves some of the operations of the U.S. Coast Guard (CG). The latter will be described in a somewhat modified manner, partly for the sake of an easier explanation and partly for reasons of proprietary information. Further, it is not completely certain that the solution presented here will be the final one we arrive at in our later studies or the one accepted by the CG.

The (hypothetical) *command hierarchy* of CG can be presented as shown in Fig. 6.4.1.

We have to consider the activities and responsibilities at the different levels of the hierarchy. The major tasks at the *Station level* are:

- Aids to Navigation — ranging from navigational charts to lighted buoys;
- Search and Rescue — helping individuals and vessels in distress;
- Enforcement of Laws and Treaties — making sure that the citizens of the United States and other countries abide by rules and regulations as prescribed by the laws of the country and international agreements.

The Station activities can be either scheduled (e.g., buoy tending, harbor patrolling) or unscheduled (e.g., search and rescue, interception, urgent repair).

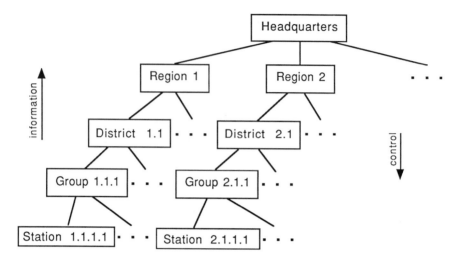

Fig. 6.4.1. — The command hierarchy of the U.S. Coast Guard

At the *Group level and higher*, the following activities are performed:

- The existing resources have to be assigned to individual Stations on the basis of historical records and anticipated future needs.
- Complex and resource-intensive tasks, requiring high-level coordination between several Stations and possibly also political decisions, have to be managed in a timely fashion.
- Relevant historical data have to be collected and evaluated for long-term planning, report generation and internal management.
- The messages between different decision making entities have to be recorded and managed (concerning destination, pathway selection, level of secrecy, level of detail, level of urgency, etc.).
- A proper information base has to be maintained about navigational details, current resources and manpower, organizational units, task types, expected scenarios, legal procedures, and the like.

A partial list of the activities performed at the *Headquarters   level* contains the following:

- To represent the interests, needs and concerns of the Coast Guard toward the Administration and the Congress;
- To request and justify budget allocations;

- To manage the Vehicle Acquisition Problem — that is, to spend the budget of the CG in the most cost-effective manner.
- To manage individual CG programs, such as Aids to Navigation, Search and Rescue, Enforcement of Laws and Treaties, etc.

6.4.3. *The Approach*

Our design is based again on a Distributed Planning and Problem Solving system. Each decision making entity in the hierarchy corresponds to a node, equipped with a processor, in a computer network. The node serves as the local unit's Command, Control and Communication Center (C^4). Each node is connected for communication purposes with its hierarchically superior node and the geographically adjacent nodes. (The term 'adjacent' is meant only symbolically. Communication is usually carried on by radio links and broadcasting can be "at large" or aimed at a selected subgroup of nodes or at an individual single node.)

Emphasis in this approach is placed on computation rather than on communication since, broadly speaking, the latter is more expensive, error-prone and has a bandwidth-limited speed. Further, the scrambling and unscrambling of sensitive messages is cumbersome and expensive under field conditions.

Each Station node can be represented schematically as in Fig. 6.4.2. There are distinct roles for each module:

The *Knowledge Base*

- maintains information about navigational details and status of the environment within the Station's jurisdiction, its own vehicles, their properties and usage statistics, crew profiles, personnel information, tasks, current and past task statistics;
- provides access mechanism for adding, changing, deleting and displaying records.

The *Plan Generator*

- invokes an expert system to retrieve qualitative ("skeleton") plans for given tasks;
- selects an optimum resource-mix (vehicles) for given tasks;

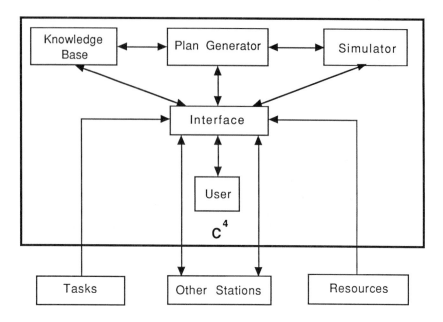

Fig. 6.4.2. — The node-based Command, Control and Communication Center

- cooperates with the Simulator to refine and validate plans and to obtain expected performance statistics;
- presents plans to the user in tabulated and graphical forms, together with its expected performance assessment;
- enables the user to select one of the plans or to over-ride all system-generated plans with his own;
- automates also the planning and scheduling of all routine operations (e.g., preventive maintenance).

The *Simulator*

- keeps the status of tasks and resources up-to-date (using both time- and event-driven simulation);
- simulates the execution of plans generated, evaluates them quantitatively, and informs the Interface about the status of the world to be expected;
- collects information about past tasks and resources;
- can make up hypothetical scenarios to expand the knowledge base of the Plan Generator;

- enables the user to monitor the progress of task execution via tabulated and graphical displays.

The *Interface*

- forwards user requests to the relevant module;
- receives information about tasks and sends instructions to resource owners;
- provides graphical and tabulated display of the current status and of the progress of real and simulated tasks for the user;
- maintains Knowledge Base coherence for consistency and accuracy;
- times appropriately various concurrent activities of the system;
- coordinates Station-to-Station and Station-to-Group communication.

The programming modules at the node level are shown in Fig. 6.4.3. Their roles are as follows:

- The *User Interface* serves as a communication link between the user and the system;

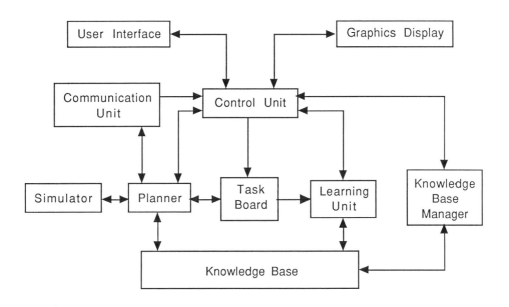

Fig. 6.4.3. — Programming modules at the node level

- The *Graphics Display* shows the area of jurisdiction within the current scope (see the definition of scope later); it also enables the user to view simulated or actual plan executions with shifting and zooming possible;
- The *Communication Unit* serves as an interface with other nodes;
- The *Control Unit* channels I/O information to correct destination, identifies and activates operational units, checks for job completion, and provides information feed-back of its quality;
- The *Simulator* is used by the Planner to test and evaluate plan quality, and to compare expected and actual situations;
- The *Planner* generates plans, takes care of dynamic plan modification, and monitors plan execution;
- The *Task Board* handles the task assignment between nodes, deals with multiple and simultaneous tasks, and continually records task execution;
- The *Learning Unit* is activated after each plan execution to modify the information and the processes used by the Planner on the basis of experience and user suggestions;
- The *Knowledge Base Manager* takes care of the creation and modification of the Knowledge Base and maintains its consistency;
- The *Knowledge Base* contains information pertaining the planning process.

6.4.4. *The Activities at Different Levels*

The proposed system has to perform the following sequences of *activities* at the *Station and Group levels* :

- A *task* with proper characterization arrives at the Station in whose area of jurisdiction it occurs. The Station becomes the *Coordinator* for that task.
- A *script-based expert system* in the Knowledge Base is activated to produce the required *resource-mix* for the task at hand.
- The Coordinator identifies different alternative *source-mixes* from which the necessary resources may be available. Such can be idle resources available at the Station, resources assigned to lower priority tasks within the Station's jurisdiction or, as a last choice, resources belonging to adjacent Stations. The last alternative uses the concept of *dynamic scoping* which needs to be explained next.

Figure 6.4.4 shows a hypothetical coast line along which several Stations are located. A resource-intensive task is reported within the jurisdiction of, say, Station *N*. If this Station cannot take care of the task on its own, a request for *extended scope* is sent electronically to the adjacent Stations *N-1* and *N+1*. The respective processors there consider their available resources and, if sufficient, *re-assign* them (temporarily) to Station *N*. The appropriate information is then shared by all three Stations. However, if their *joint resources* available are still not sufficient, Stations *N-1* and *N+1* request extended scope of Stations *N-2* and *N+2*, respectively. This process goes on until enough resources are re-assigned to Station *N*. In turn, as tasks become accomplished, resources from other Stations return to their original "owner" — hence the name 'dynamic scoping'.

We also note that the *constraint of limited resources* is aggravated by vehicle breakdowns, their scheduled and unscheduled maintenance, personnel availability, and different levels of resource preparedness. The management of too complex or resource-intensive tasks may go up to the *Group level* or

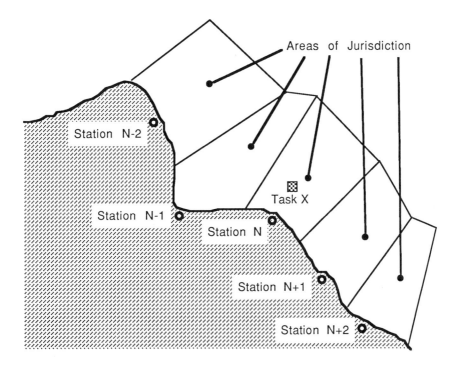

Fig. 6.4.4. — The idea of dynamic scoping

even higher when the above-described technique of dynamic scoping is over-ridden by the human operator for some particular reasons.

- The Plan Generator submits the *tentative plans* with different source-mixes to the Simulator for quantitative and qualitative evaluation.
- Tentative plan executions are *monitored* by the User and ordered by the system according to their *utility*. The evaluated plans, and the basis of their evaluations, are then presented to the User.
- The User *accepts* one from the list of alternate decisions or *over-rides all* with his own plan.
- A *learning mechanism* gradually improves the Knowledge Base and the planning process.
- *Interrupt mechanisms* at different levels can take care of emergency situations.
- *Statistical data collection* serves the purposes of the learning mechanism and report generation.

The *activities* that need to be performed at the *Group levels and above* include the following:

- On the basis of the historical records collected and the scenarios generated in using task-mixes extrapolated to the future, an *optimum re-allocation of existing resources* to individual stations has to be made.
- The *management of complex and resource-intensive tasks*, requiring high-level coordination and political decisions, are made by the combined man-machine system — leaving the ultimate decision making in human hands. *"What-if" type questions* are answered reliably by the system, which facility provides a powerful decision support for the user.
- Relevant historical data are collected and evaluated for *long-term planning*, *report generation* and *internal management*.
- The *management of messages* is automated concerning destination (whether it is broadcasting at large, or aimed at a group of nodes selected by need or individually addressed nodes), pathway selection (default or alternative pathways), level of secrecy (encoding and decoding), level of detail (instructions, reports at different degrees of abstraction), and level of urgency.
- The Knowledge Base must also be automatically updated with reference to new or modified Stations and resources, organizational units, task types, expected scenarios, constraints, and other relevant changing information.

Finally, we show how the system would contribute to the solution of the *Vehicle Acquisition Problem*, which is one of the important activities performed at the *Headquarters level*. There are two alternative ways of formulating the Vehicle Acquisition Problem:

(1) Within a given budget, what set of resources should be purchased to meet the needs in the best way?

(2) To accomplish a given set of tasks, what set of resources should be purchased, subject to the constraint of minimum expenditure?

The *sequence of activities* is as follows:

- Generate a historically valid extrapolation of *time sequences of task-mix scenarios*;
- Generate realistically-sized and systematically representative choices of feasible *resource-mixes*;
- Run the *Plan Generator* in conjunction with the *Simulator* under optimum operating conditions (with reference to resource allocation to tasks over space and time) for each meaningful *task-mix and resource-mix combination*;
- Measure the *quality of task accomplishments*, using plausible metrics (including the cost of resource amortization, cost of resource utilization, time spent with the accomplishment of all tasks, value of partial/full goal attainment, loss due to incomplete/missed goal attainment, etc.);
- *Order the feasible resource-mixes* for each task-mix, according to their quality as computed above;
- *Select the best resource-mix* within the given budget or *present the least expensive resource-mix* capable of handling all task-mixes.

6.4.5. *The Salient Features of the System*

The system under development, a Distributed Planning and Problem Solving System for allocating distributed resources to distributed and moving tasks, has a number of important features. These include the following:

- The Knowledge Base can be extended and modified either automatically, through the learning processes built into the system or manually by the user.

- Scenarios of different task-mixes and resource-mixes — satisfying a given set of criteria — can be specified either by the user in a high-level manner or by the system in a random or a systematic manner.
- New planning strategies can be defined by the user and optimized by the system.
- Collecting statistical data and evaluating mission performance are easy and inexpensive. They can be summarized under user control, for example, with reference to the current scenario, a given task, a type of tasks, a given resource, a type of resources, etc.
- The user can specify the ratio between simulated and real time between reasonable limits.
- The scope of plan generation with reference to resources and stations is need-driven and changes dynamically between and even within tasks.
- The same program can be used for plan generation and problem solving at every station — only a part of the knowledge base characterizing, for example, the geographical environment is different.
- The graphic display assists the user in monitoring, approving or modifying system decisions;
- Incomplete and partially erroneous information can still lead to feasible and meaningfully effective plans.
- Additional and more precise information can be used by the system to amend the current plan.
- Further tasks can be accepted during the planning and even the plan execution activity.
- System responses to changing conditions (e.g., an increase in the level of urgency of a given task) are prompt due to the 'dependency-based reasoning' technique in planning, which makes use of causal relations discovered between planned action steps and goal components.
- Communication to resources and other stations is goal-oriented, unambiguous and based on the 'need-to-know' principle.
- Trade-off calculations between costs and benefits are readily available, under all relevant conditions and using a variety of scenarios and assumptions.
- In addition to being an automated, geographically distributed decision support tool, the system provides a rigorous method of solving the Vehicle Acquisition Problem.
- Finally, it needs to be emphasized that, in view of the time-scale characterizing the task environment, the ultimate decision making can be left to the human user of the system. It is, however, also true that the environment is *time-critical*, has various constraints on resources, is

subject to legal and physical limitations, and as such represents in general an *information overload* for the human decision maker.

6.5. A Distributed System for Street Traffic Light Control

6.5.1. *Introduction*

Traffic engineers have been using different tools of mathematics, statistics and computer science to devise systems that can improve the traffic conditions of our congested cities. Some techniques of Artificial Intelligence have also been employed to generate, for example, better *constant* control of street traffic lights and real-time expert systems controlling street traffic lights *centrally* in a dynamic fashion.

Distributed and dynamic control can, however, offer a number of advantages. The reasons are as follows:

- Spatially and chronologically *local conditions* are usually more relevant to the decisions to be made. Traffic accidents, the ending of a major sport event, changes due to road repair or an "unscheduled" holiday are examples of local changes that cannot be considered by any centrally controlled regime.
- The *rate of change* in local conditions is usually very high. Even if high-performance sensors are available for data input, communication and computational bottlenecks would not allow the existence of a timely and responsive control environment.
- A centrally organized, real-time planning technique of satisfactory quality is not feasible because of the *overwhelming amount of data* to be processed and the *large number of decisions* to be made and communicated to the traffic lights.
- Changes to a distributed control system are easy and inexpensive to make, when the 'permanent' traffic environment changes.

6.5.2. *The Approach*

We have adopted the following working conditions and made the following assumptions for the project under development:

- The *display subsystem* and the *traffic simulation subsystem* connected to the Plan Generator component are based based on our 'Roads Scholar' program, originally written for the Advice Taker/Inquirer project.
- The *street configuration* is according to the 'Manhattan grid' (see Fig. 6.5.1). It means that left turns are prohibited, all streets are two-way, cross at right angles, and run either in the North-South or East-West direction.
- The resulting system should be *modular* so that the addition of more realistic features (e.g., one-way streets, multi-lane tunnels and bridges with changeable lane directions, and the like) should cause no significant reprogramming to be done.
- There is *one processor at each intersection,* which communicates *directly* with the four processors at the adjacent intersections.
- The *communicated information* is three-fold:
  - *raw data* (essentially, the number and the speed of cars going in each of the four directions at an intersection),
  - *processed data* (the type and the rate of change of certain traffic flow features),
  - *expert advice* (e.g., "lengthen the period of green light in the East-West direction").

It should be noted that the latter two categories of information can propagate over an indefinite number of intersections but with gradually changing contents. Such "combined" information coming from many intersections along a given direction is the weighted average of the contributing information — the farther away the source, the less important its contribution is.

- The operation of the whole system is based on
  - a set of collaborating real-time expert systems which work in conjunction with a simulation-based planning system,
  - a limited amount of noisefree communication which triggers both gradual and sudden changes in the traffic light control regime.
- There are several possible *criteria of operation*. The 'objective function' to be minimized can be the average travel time, the maximum waiting time at intersections, or the average number of stops during travel, etc.
- We have intended to employ two types of rules in the expert system:
  - *traffic policeman's rules* — based on our own introspection and information obtained by interviewing others,
  - *mathematician's rules* — based on information gleaned from articles and expert opinion.

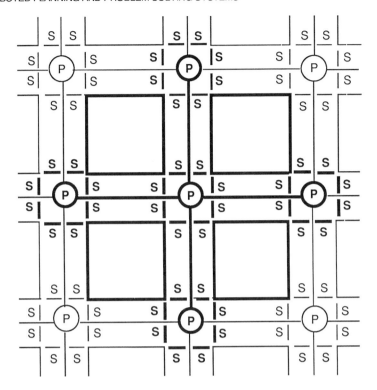

Fig. 6.5.1. — The "Manhattan grid" used as the street pattern. There is a processor, *P*, at every intersection, which receives input data from its own eight sensors, *S*, and various types of information from the processors at the adjacent four intersections

- It has seemed a good idea to *evaluate individual rules*. Various measures of rule performance are to be tried, such as a rule's relative contribution to the improvement of system performance, number of times a rule fires, the rule's computational cost and degree of robustness, etc.
- We have intended to introduce *automatic learning features* in the rule base. The dimensions of learning can be: adding/deleting rules, changing rule parameters, changing rule priorities for the resolution of conflict between competing rules, etc.

We hope to be able to show that the system would ultimately produce several benefits, such as faster traffic flow, more efficient usage of available roads, reduced air pollution, reduced fuel consumption, fewer accidents, and lower driver frustration.

6.5.3. *The Control Strategies*

There are different strategies for the control regime — each with its own set of rules to follow. In describing these, we will use the average waiting time of cars going though an intersection as the objective function to be minimized. (The terms 'intersection' and 'traffic light' will be used interchangeably.)

The control variables for the traffic light are:

- the length of the cycle;
- the length of "active time" (sum of the periods of yellow[1] and green lights in one or the other direction);
- the point of time when the cycle starts.

There are three basic strategies; the third one can be divided into three further categories, depending on which type of control variable is used. We present the strategies in the order of usage priority.

Strategy I: *The Semiactuated Regime*

This strategy is to be used when the traffic flow in one of the intersecting streets is extremely small. The light stays green in the busier direction until one or more cars approach the intersection in the less busy street. Then the light turns green in the latter direction as needed, up to a predetermined maximum time.

Strategy II: *The Platooning Regime*

In moderate traffic, it is a good idea to encourage cars to travel in "platoons", in small groups separated by gaps. Ideally, the light should be green when the

---

[1]Traffic experts do not vary the time period for the yellow light, t_y but keep it at about

$$t_y[sec] = 0.1 * v_{max}[m/h]$$

where v_{max} is the speed limit.

6. DISTRIBUTED PLANNING AND PROBLEM SOLVING SYSTEMS

cars are coming to cross the intersection, and red when the gaps appear. The staggering of the traffic lights to this effect is easily controlled by a static and global control system. However, distributed systems can also accommodate platoons.

Since it is more effective to stagger the traffic lights to handle platoons than to follow one of the lower priority schemes to be described below, the platooning scheme should be followed when possible.

Strategy III: *The Regime To Control Individual Characteristics Separately*

This mode of operation is based on three sets of rules which are ranked in the order of usage priority. Each controls a different variable.

Substrategy IIIa: *Modify Cycle Length*

As a general heuristic, it has been found that when the traffic flow is heavy (say, above 1300 cars/lane/hour), longer cycle lengths speed traffic. In turn, when traffic is lighter, shorter cycle lengths are advisable. However, cycles of longer than 180 seconds or shorter than 40 seconds are inefficient and should not be used.

Substrategy IIIb: *Change Cycle Splits*

The cycle should, in general, be split so that the direction with heavier traffic flow receives the longer green light.

Substrategy IIIc: *Change Cycle Start Time*

If it is found that too high a proportion of cars arrive on the red light or they have to wait longer than seems appropriate with the given the flow and cycle split, the cycle start time should be adjusted to reduce waiting time. Depending on the prevailing traffic pattern, such a measure may be of long-term help or may improve the situation only temporarily.

6.5.4. *The Information Communicated between Controllers*

The following type of symbolic and numerical information is the result of some calculation on locally sensed data, which must then be transmitted to the appropriate adjacent processor:

- a congestion is being experienced at the controlled intersection, which is moving toward the adjacent one;
- a congestion is moving from the adjacent intersection toward the controlled one;
- a severe congestion is being experienced at the controlled intersection, which is moving toward the adjacent one;
- a severe congestion is moving from the adjacent intersection toward the controlled one;
- data on cycle length;
- data on cycle start time;
- data on cycle split time;
- a car crossed the intersection in the direction at hand when the light turned green;
- the number of cars having crossed the intersection in the direction in question.

In the statements used so far, *flow* is the number of cars clearing the intersection per minute and per lane, *congestion* occurs when a car must wait a whole cycle before clearing the intersection, and *severe congestion* occurs when a car must wait at least two cycles before clearing the intersection.

6.5.5. *Some Notation and Definitions To Be Used with the Rules*

First, it should be noted that the parameters in the rule expressions are local, unless indicated otherwise, and will be varied by the learning components of the program in order to optimize the traffic flow.

The following notation is introduced:

P: a period of time;

dir: any of the four directions, N, S, E or W — dir_1, being any one of the above, is followed by dir_2, dir_3 and dir_4, each rotated clockwise by 90°; further,

NS : the North and South direction;
EW : the East and West direction;

$suc(i, dir)$:   the successor of traffic light i in direction dir;

$pred(i, dir)$:   the predecessor of traffic light i in direction dir;

L_i: the cycle length of traffic light i;

Z_i: the start time of the cycle (the point of time in the cycle when the light goes green in the NS direction);

H_i: the change time of the cycle (the point of time in the cycle when the light goes green in the EW direction);

$G_i(dir)$:   the time period when the traffic light is active (i.e., green or yellow in the direction dir);

$$H_i = Z_i + G_i(NS) \bmod L_i;$$

$S_i = G_i(NS)/L_i$: the split (the fraction of cycle time for traffic light i in the NS direction);

$R_i(dir, P)$:   the number of cars per lane travelling through the intersection i in direction dir in time period P;

$SR_i(dir, P)$:   the number of cars per lane crossing intersection i in direction dir in time period P,

$$SR_i (NS, P) = R_i (N, P) + R_i (S, P),$$
$$SR_i (EW, P) = R_i (E, P) + R_i (W, P);$$

$A_i(P)$:   the activity at intersection i in time period P,

$$A_i (P) = SR_i(NS, P) + SR_i (EW, P)$$
$$= R_i(N, P) + R_i (S, P) + R_i (E, P) + R_i(W, P);$$

C_j: the j-th car;

$W_i(C_i)$:   the waiting time of car C_j at intersection i;

$TW_i(P)$:   the total waiting time at intersection i in time period P,

$$TW_i(P) = \sum_{j=1}^{n_i} W_i (C_i),$$

where n_i is the number of cars that arrive at the intersection i in time period P;

$AW_i(P)$:   average waiting time at intersection i in time period P,

$$AW_i(P) = \frac{\sum_{j=1}^{n_i} W_i(C_i)}{n_i},$$

where n_i is the number of cars that arrive at the intersection i in time period P;

$TWS_i(dir, P)$:   total waiting time at intersection i in direction dir in time period P;

$AWS_i(dir, P)$:   average waiting time at intersection i in the direction dir in time period P;

t: time in seconds;

t_c: current time;

$B(i, j)$:   average time it takes the first car stopped at traffic light i to get to traffic light j, after traffic light i turns green;

$GB(i, j)$:   average time it takes a car coming through the green and uncongested traffic light i to reach traffic light j;

$C_i(dir)$:   there is a congestion in direction dir at intersection i if

$$AWS_i(dir, L_i) > L_i;$$

$SC_i(dir)$:   there is a severe congestion in direction dir at intersection i if $AWS_i(dir, L_i) > X^*L_i$. (X is set for 2, as specified later in the Definitions.)

As we have stated before, the main variables L_i, Z_i and S_i are controlled by the processor on their own and independently of each other. Other variables, such as H_i and $G_i(dir)$, are functions of these main variables but are included in the rules to improve clarity.

6.5.6.   *The Rules to Control the Traffic Light Regime*

We now list some representative rules in the five categories specified in Section 6.5.3. The rules are first described in English, then expressed in mathematical form in using the definitions and the notation given before.

A. *A Rule That Sets Up Semiactuation (Strategy I):*

> **if** *traffic flow* on one street is greater than X
> **and** on the perpendicular street is less than Y
> **then** switch to *semiactuation mode* :

$$(SR_i(dir_1, P) < X) \wedge (SR_i(dir_2, P) > Y \longrightarrow$$
$$\text{switch to semiactuation mode.}$$

B. *Rules That Set Up Platooning (Strategy II):*

> **if** there is no *congestion* in any *direction* at an intersection
> **and** the *traffic flow* in the direction dir_1 is X times greater than the *traffic flow* in direction dir_2
> **and** the *preceding* and *succeeding traffic lights* turned green in direction dir_1 within the last Y seconds
> **then** set Z (the time point when light turns green in direction dir_1) to be equal to B (the time point when the first car should arrive from one of the dir_1 *traffic lights*) :

$$\neg C_i(dir_i) \wedge \neg C_i(dir_2) \wedge \neg C_i(dir_3) \wedge \neg C_i(dir_4) \wedge$$
$$(SR_i(dir_1, P) > X^*SR_i(dir_2, P)) \wedge$$
$$(((t_c \bmod L_{suc(i)} - Z_{suc(i)}) < Y) \wedge$$
$$(((t_c \bmod L_{pred(i)}) - Z_{pred(i)}) < Y) \longrightarrow$$
$$Z_i := \min\{(Z_{suc(i)} + B_{suc(i)}) \bmod L_i),$$
$$(Z_{pred(i)} + B_{pred(i),i}) \bmod L_i)\}.$$

> **if** the *traffic flow* in one *direction* is X times greater than the sum of the *traffic flows* in the other three *directions*
> **then** set the traffic light so that it turns green in the busier *direction* at B (the time point when the first car arrives from the *preceding traffic light*) :

$$R_i(dir_1, P) > X^*((R_i(dir_2, P) + R_i(dir_3, P) + R_i(dir_4, P))) \longrightarrow$$
$$Z_i := min\{(Z_{suc(i)} + B(suc(i),i)) \bmod L_i),$$
$$(H_{suc(i)} + B(suc(i),i)) \bmod L_i)\}.$$

C. *Rules That Set Cycle Length (Substrategy IIIa):*

if the *average waiting time* in all four *directions* is greater than *V* times the *cycle length*
and the *cycle length* is less than 180 seconds
then lengthen the *cycle length* by *X* where *X* is *Z* times the *average waiting time* at the *traffic light* but in no case greater than *Y* or 180 seconds, whichever is smaller:

$$(AWS_i(dir_1) > V^*L_i) \wedge (AWS_i(dir_2) > V^*L_i) \wedge$$
$$(AWS_i(dir_3) > V^*L_i) \wedge (AWS_i(dir_4) > V^*L_i) \wedge$$
$$(L_i < 180 \ sec) \longrightarrow$$
$$L_i := min\{180, L_i + min\{Y, Z^*AW_i(P)\}\}$$

---

if there is no *congestion* in any *direction* **and** the *cycle length* is greater than 40 seconds
  and the *total flow* is less than X_1
    and the *delay* is less than Y_1
      and if the *cycle length* plus Z_1 is less than 40 seconds
      then set *cycle length* to 40 seconds
      else shorten the *cycle length* by Z_1;

  else if the *total flow* is less than X_2
    and the *delay* is less than Y_2
      and if the *cycle length* plus Z_2 is less than 40 seconds
      then set *cycle length* to 40 seconds
      else shorten the *cycle length* by Z_2;

  else if the *total flow* is less than X_3
    and the *delay* is less than Y_3
      and if the *cycle length* plus Z_3 is than 40 seconds
      then set *cycle length* to 40 seconds
      else shorten the *cycle length* by Z_3;

[Here in all $X_1 < X_2 < X_3$, $Y_1 < Y_2 < Y_3$, $Z_1 < Z_2 < Z_3$]:

$$\neg[C_i(dir_1)]\wedge \neg[C_i(dir_2)]\wedge \neg[C_i(dir_3)]\wedge \neg[C_i(dir_4)]\wedge (L_i >40)\wedge$$
$$(A_i(P) < X_1{}^*P)\wedge(AW_i(P) > Y_1)\wedge$$
$$[L_i + Z_1 < 40]\longrightarrow L_i := 40;$$
$$\textbf{else}\qquad L_i := L_i - Z_1;$$

$$\textbf{else}\qquad (A_i(P) < X_2{}^*P)\wedge(AW_i (P) > Y_2)\wedge$$
$$L_i + Z_2 < 40 \longrightarrow L_i := 40;$$
$$\textbf{else}\qquad L_i := L_i - Z_2;$$

$$\textbf{else}\qquad (A_i(P) < X_3{}^*P) \wedge(AW_i(P) > Y_3) \wedge$$
$$L_i + Z_3 < 40 \longrightarrow L_i := 40;$$
$$\textbf{else}\qquad L_i := L_i - Z_3.$$

D. *Rules That Set Cycle Splits (Substrategy IIIb):*

The following rules must be constrained so that a meaningful proportion of the total cycle time must be given to both directions. Also, some of the rules will become redundant or never used (to be shown by the learning program).

if only one *street* is *congested* at an intersection
and that *street* is not *severely congested* at the *preceding* and *succeeding* *traffic lights*
then lengthen the *cycle spit* of the green light by X in the *direction* that is *congested* :

$$C_i(dir_1)\wedge \neg C_i(dir_2)\wedge \neg SC_{(suc(i,dir_1)}\wedge \neg SC_{pred(i,dir_1)} \longrightarrow$$
$$G_i(dir_1) := G_i(dir_1) + X$$

———————

if one *street* at an intersection is *congested* in both *directions*
and the other *street* is not *congested* in either *direction*
then lengthen the green light on the congested *street* by X :

$$C_i(dir_1)\wedge C_i(dir_3)\wedge \neg C_i(dir_2)\wedge \neg C_i(dir_4) \longrightarrow$$
$$G_i(dir_1) := G_i(dir_1) + X$$

———————

if the *total waiting time* and the *average waiting time* of cars arriving at an intersection on one *street* are greater than the same on the other *street*
then lengthen the green light on the first *street* by X :

$$(TWS_i(dir_1) > TWS_i(dir_2)) \wedge AWS_i(dir_1) > AWS_i(dir_2)) \longrightarrow$$
$$G_i(dir_1) := G_i(dir_1) + X$$

---

if there is *severe   congestion* in both *directions* on one street at an intersection

and no *congestion* on the other *street*

and there is no *severe   congestion* at the *preceding* and *succeeding traffic lights*

then lengthen the green in the first *street* by X :

$$SC_i(dir_1) \wedge SC_i(dir_3) \wedge \neg \dot{C}_i(dir_2) \wedge \neg C_i(dir_4) \wedge$$
$$\neg SC_{suc(i,dir_1)} \wedge \neg SC_{pred(i,dir_1)} \longrightarrow$$
$$G_i (dir_1) := G_i (dir_1) + X$$

---

if there is *severe   congestion* or *congestion* in both *directions* and both *streets* at an intersection

and there is *severe   congestion* on a *succeeding   intersection* on one *street* but not on the other

then lengthen the green light on the *street* with clear *succeeding intersection* by X :

$$(C_i(dir_1) \vee SC_i(dir_1)) \wedge (C_i(dir_2) \vee SC_i(dir_2)) \wedge$$
$$(C_i(dir_3) \vee SC_i(dir_3)) \wedge (C_i(dir_4) \vee SC_i(dir_4)) \wedge$$
$$SC_{suc(i,dir_1)} \wedge SC_{suc(i,dir_3)} \wedge \neg SC_{suc(i,dir_2)} \wedge \neg SC_{suc(i,dir_4)} \longrightarrow$$
$$G_i(dir_2) := G_i(dir_2) + X$$

---

if there is no *congestion* in any *direction* at an intersection

and the *traffic flow* on both *streets* is greater than X

and the *average waiting time* in one *direction* is Y times greater than the same in the other

then change the *traffic light* to green in the first *direction* V seconds earlier

and increase the *length* of the green light in the same direction by V seconds:

$$\neg C_i(dir_1) \wedge \neg C_i(dir_2) \wedge \neg C_i(dir_3) \wedge \neg C_i(dir_4) \wedge$$
$$(SR_i(dir_1) > X) \wedge (SR_i(dir_2) > X) \wedge$$
$$AWS_i(dir_1,P) = Y^*AWS_i(dir_2,P) \longrightarrow$$
$$Z_i := Z_i - V, \quad G_i(dir_1) := G_i(dir_1) + V$$

---

if one *street* at an intersection is *congested*
and the other *street* is not
then increase the *length* of the green light in the *congested direction* by X :

$$C_i(dir_1) \wedge \neg C_i(dir_2) \longrightarrow G_i(dir_1) := G_i(dir_1) + X$$

E. *A Rule That Sets Cycle Start Time (Substrategy IIIc):*

if the *average traffic flow* crossing an intersection is greater than X
and the *average waiting time* is greater than Y times the one that would be
expected given the *traffic flow*
then start the cycle V seconds earlier

$$A_i(P) > X) \wedge (AW_i(P) > Y^*A_i(P) \longrightarrow Z_i := Z_i - V$$

6.5.7. *On Scenario Generation*

A large number of experiments need to be performed in trying to optimize the
rule base of the *distributed, cooperative expert systems*. Each series of
experiments will have to be provided with an overall traffic pattern that
applies to a *characteristic period* of the day (e.g., early morning rush hour,
mid-day traffic, late afternoon rush hour, evening traffic and night traffic),
of the day of the week (workday, Saturday, Sunday, other holiday) and,
possibly, of the season of the year.

Recall that we have assumed an indefinitely large area of the Manhattan
grid type. The following idea enables us to study a relatively small segment
of it only, without risking unrealistic traffic situations. Let us call the *area
of concern* a rectangle cut out from an indefinitely large *Manhattan grid*. If
the number of intersections in the E-W direction is w (width) and in the N-S
direction d (depth), we can number all the intersections within the area of
concern as shown in Fig. 6.5.2.

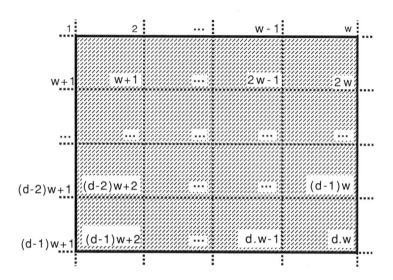

Fig. 6.5.2. — The *area of concern* cut out of an indefinitely large Manhattan grid; the four *peripheries* bordering it are marked by heavy lines and the intersections are numbered where *w* is the width of the area of concern and *d* is its depth — both in 'number-of-intersections' units

Further, let the area of concern be surrounded by four *peripheries*, each of which contains a sequence of street intersections along the streets that border it. Therefore, the numbers of the intersections along the four peripheries are as follows:

Top, E-W direction:	1	2	...	w
Left, N-S direction:	1	w+1	...	(d-1)w+1
Bottom, E-W direction:	(d-1)w+1	(d-1)w+2	...	d.w
Right, N-S direction:	w	2w	...	d.w

We can represent the traffic pattern to be generated for one characteristic period defined above by two *wxd* matrices. The elements stand for the "source" and "sink" specifications, respectively, for each intersection — that is, the number of cars originating from a given intersection and coming to it as a destination. The actual values of the elements, produced by pseudo-random number generators, have two types of constraint:

- The sum of the source numbers equals the sum of the sink numbers, and they both equal a constant representing the characteristic period.
- The source and the sink numbers associated with the intersections on the peripheries equal a user-specified constant times the respective numbers obtained. This way, we can take care of the fact that a lot of traffic goes to and comes from the area of concern across the peripheries.

6.5.8. *The Optimization of the Rule Base*

The following approach is adopted for the optimization of the rule base. First, there is a possible natural segmentation of the rule base. Each of the strategies and substrategies listed in Section 6.5.3 represents a particular mode of operation whose controlling rules belong to a distinct segment. We have identified a small set of *meta-rules* that, in response to the current traffic pattern and characteristic period, point to the applicable rule segment for further processing.

Next, we define the *average travel time* of a car, calculated over all trips during a characteristic period, as the basis of comparison between different rule bases and the *quality measure* of a given rule base. (Rule bases may differ in their membership of constituent rules and/or in the parametric values contained in individual rules.) Let us call the *core set of rules* of a given segment those rules that are indispensable for the control of operation in the segment, regardless of the quality of operation. The ordering of rules in the core set, and in any later extended set, is always according to the frequency of rule usage (starting first with a random ordering). The search for an applicable rule in a segment begins at the top, and the first rule matching the current conditions is fired — a simple and inexpensive way of dealing with the problem of conflict resolution.

We then *optimize* the parameters of the rules in the core set through a usual hill-climbing technique. In the further computations, these parameters of the rules in the core set are kept constant. The next task is to decide which of the additional rules outside the core set should augment the corresponding rule segment. We add one rule at a time, optimize its parameters, and measure its contribution to the quality of operation. We leave only those rules in the final augmented segment whose contribution is above a certain threshold value.

6.6. Summary

After a general introduction to Distributed Planning and Problem Solving, we have discussed four areas of application, the latter two of which are still under development. The four areas are very different in regard to their

- domains,
- reliability concerns,
- quality measures,
- computational and communication requirements,
- timing aspects, etc.

There are, however, certain common features they share, such as

- they all have geographically distributed input and output operations,
- each node cooperates with a selected set of other nodes,
- the subgroup formation (partitioning) of the nodes may change according to task status, current node activity and availability,
- each system has a need for reliable and gracefully degrading performance when some operational and/or computational units become disabled,
- some knowledge is universally needed by all participating decision making entities in the system but some is specific to individual nodes,
- the planning process has to provide a satisfactory schedule and resource allocation *in time* because the subsequent plan execution must meet certain chronological constraints.

We have seen that planning is an important element in widely different areas of computing activity. *Plan generation* involves reasoning about timing future actions and estimating their consequences. The consequences must be consistent with given goals and constraints. The search involved in the planning process must be controlled to avoid combinatorial explosions. *Plan execution* is aggravated by the uncertain (incomplete, changing and possibly inconsistent) knowledge used with plan generation. Errors must be discovered in time and rectified.

Some of the methodology described has a fairly general applicability but, at this stage of development, many techniques are still designed only for a particular application. This fact, among others, indicates a significant need for further studies on planning.

6.7.   Acknowledgements

The following people have worked on the individual projects: Ron Lo on Air Traffic Control, Qing Ge and Ji Gao on the Distributed System for Manufacturing Control, Uttam Sengupta and Cem Bozsahin on the System for Distributed Resources and Tasks, and John Stapp on the Distributed System for Street Light Control.

7. Causal Modelling Systems (CMS and NEXUS)

7.1. Introduction and Perspectives on Causation

At the risk of oversimplifying concepts and terms that are important to philosophers, the elements of processes involved in scientific inquiry — observational procedures, sequences of argument, methods of representation and computation, concerns about validation — basically aim at discovering *causal relations* between patterns of empirical phenomena. On the basis of these causal relations, science has the goal of constructing concise and systematically organized *theories* that describe and explain the world of nature. Technology then makes use of the results of the scientific enterprise to build reliable, effective and economically justifiable products to serve given needs and objectives.

Our goal has been to make a few steps toward automating the process of *hypothesizing* and *identifying causal relations* and utilizing them in specific applications. Before we discuss the related efforts and their results, it is in order to explore briefly the relationship between different disciplines and the concept of causality.

7.1.1. *The Philosophical View of Causality*

The problem of causality has been discussed in philosophy since pre-Socratic philosophers [22] and the concept itself has assumed a central role in sciences ever since Aristotle's treatise of the subject. He defined four basic causes: the material cause (what the entity is made of), the formal cause (the pattern that the changing entity acquires), the efficient cause (what makes the change to occur), and the final cause (the goal toward which the change aims). The efficient cause has been the epitome of scientific explanations and, combined with the final cause, has also survived in social sciences.

At the same time as Rationalism developed on the European continent, culminating in Descartes' objective of constructing self-consistent and deductive logical systems, the British philosophers took a different approach. Francis Bacon, author of the method of exhaustive induction, emphasized empirically observed facts and sensory perception. David Hume stated that experience alone could never reveal the relationship between cause and effect, and the latter cannot be inferred by reasoning about our sensory impressions. Causality, according to him, is a natural disposition of humans to categorize events; it simply shows our mental character and the way we experience nature. Hume defined causation as a phenomenon which exists when three conditions hold: (a) cause and effect are contiguous in time and space; (b) cause precedes effect in time; (c) there exist a constant conjunction between cause and effect. The first two conditions are intuitive notions, although they are also brought into question in reference to delayed effects and the relativistic view of the universe, respectively. Nevertheless, the third condition has been the one which has spawned off various accounts of the concept in the literature.

Constant conjunction was defined as "... an object followed by another, and where all the objects, similar to the first, are followed by objects similar to the second". A useful interpretation of the third condition is that causation is the process of detecting regularities between instances of events. This view is exemplified by the statistical and probabilistic accounts of causation.

Until the mid-18th century, the notion of causality was so fundamental to all scientific explanations that scientists had made no distinction between causal and non-causal laws. However, Hume's critique of causation has eventually led to better founded definitions later on. Another outcome of his views is the recognition of the difficulty of capturing different facets of causality under a single framework. Physical scientists emphasize spatio-temporal properties of events as the primary focus of causal analysis. Social scientists concentrate on finding causes of changes in social inertia and are interested in goal-directed behavior and its relation to causal attribution. Psychologists are concerned about the perception of causality.

Since causation has been a very popular topic in modern philosophy, an even modest coverage of related studies is beyond the scope of this work. We note, in the next section on mathematically oriented studies, a few recent contributions of a philosophical flavor which have been influential in the design of some computational causal models and causal heuristics.

7.1.2. *The Probabilistic, Statistical and Logical Approaches to Causality*

In probabilistic formulations, causation is defined as a conditional probability. For instance, Suppes [212] defines A as the *prima facie* cause of B when

- $prob(A_t) > 0$,
- $prob(B_{t'}|A_t) > P(B_{t'})$,
- $t < t'$

where A_t and $B_{t'}$ are events at time points t and t', respectively. In this treatment, Suppes also defines a way of differentiating the genuine causes from spurious ones which may be included among prima facie causes.

Another probabilistic view by Good [89] leads to a "causal calculus" for finding the degree of *causal relatedness* between events. The problem is defined as finding the *causal support* for the effect provided by cause, denoted as $Q(B:A)$. In addition to conditional probabilities of individual event's occurrence, two ad hoc metrics are specified, the strength of the (most direct) causal chain from A to B, and the strength of a "causal net" joining A to B — essentially chains alternative to the most direct one — which together are instrumental in defining the degree of causal relatedness between A and B. Good's formalism is the first one to differentiate between a causal chain and a causal net, thus accounting for the different, alternate ways of causation between two event classes. One can use these probabilistic concepts in doing extrapolations to the future or in examining an uncertain past.

Salmon's *statistical-relevance* theory [192, 193], prompted by his dissatisfaction with the deductive models of scientific explanations, is an attempt to derive causal explanations from relations of statistical relevance holding among the events. The cause of an event is found when it is in a set of events statistically related to its occurrence via a set of statistically relevant factors. His view was based on continuous processes rather than basic events. The model, similarly to the analysis of variance (ANOVA) methodology, tries to extract causal knowledge solely on the basis of statistical relations. Rogers [191] also uses a continuous process model of the environment. Processes do not interact with each other until they collide in space-time when a conjunctive fork arises. The change from the non-interactive state of the processes is considered to be caused by the fork.

Computational studies of causation have heavily favored correlational approaches. A statistical causal model extracts the level at which a dependent variable is related to the independent variables and other dependent variables. The justification for statistical analysis is based on the fact that using large data sets will help detect patterns between co-occurring events with a computable degree of confidence. The primary objection to a statistical approach is that it hides the possible existence of a causal factor affecting both the dependent and independent variables. Therefore, these variables, although statistically correlated, are not necessarily causally related. Other drawbacks of correlational models are their inability to handle non-numerical information, the lack of commonsense knowledge (e.g., relating goals to their consequences), and the difficulty of collecting voluminous unbiased data that are required to make a sound statistical conclusion.

To put it in other words, the underlying problems of the statistical-probabilistic models are that of ontology and measurability: What are events? What does *prob(event)* mean as a proposition? How can one measure event probabilities? These questions, no longer of a metaphysical nature, have not been answered satisfactorily for a general class of events. A computational model of causation needs to address the following concerns:

- the *epistemological problem*: what is causal knowledge and what are its sources;
- the *ontological problem*: what kind of information, and in what form, is used as input to draw causal conclusions;
- the *scope problem*: what range of causal phenomena (e.g., social, physical) can be handled by the model;
- the *knowledge representation problem*: how is causal knowledge represented, stored and retrieved;
- the *causal reasoning problem*: given the changing world and domain knowledge, what kind of causal inferences can be made and how are they maintained;
- the *causal learning problem*: how is causal knowledge acquired, and when and how is prior causal knowledge updated;
- the *verification problem*: how is causal knowledge tested and verified.

Several *logics* have also been used for causal reasoning [48, 96, 99, 100, 129, 217]. In these systems, a set of predicates and functions axiomatize the underlying domain, and the rules of inference carry out the task of finding causal connections. Logical models have a semi-deductive flavor in the sense that the regularity in the domain is pre-specified in the domain axioms. Causal theories in logic are theorems which are proven on the basis of the axioms of the domain. The non-monotonicity of causal reasoning and the dynamic nature of domain knowledge present difficulties in designing deductive systems for causal learning and reasoning. Moreover, reasoning is clearly separated from the underlying memory structure in logic systems. This brings about an additional difficulty of relating past experiences to new ones in revising causal hypotheses.

Computationally, causal reasoning can also be a *refutation process*. It allows to sieve through and classify large amounts of knowledge by ruling out the impossible causal associations. *Causal learning* is accomplished by inducing potential causal connections as similarities between current events and precedents of these new experiences are found.

Causation has been defined by Mackie as a difference between necessity and sufficiency of causes and effects [142]. A cause, called the *INUS condition* by Mackie, is defined as an "*in*sufficient but *n*ecessary part of a condition which is itself *u*nnecessary but *s*ufficient for the result." Thus A causes B if $(A \wedge X)Y$ is an INUS condition for B, for some X and Y. Conditions A, X, and Y are required to be non-redundant to eliminate spurious causal connections. Mackie's concern is with the analysis of singular causal statements, not causal chains. The difficulty of chaining lies in the fact that causality does not depend singly on A but either $A \wedge X$ or Y. According to Mackie, his theory amounts to a form of regularity theory for causation.

Another interpretation of the third, regularity requirement of Hume, the "constant conjunction between cause and effect", was given by Lewis [136]. He elaborated on Hume's alternate definition of the constant conjunction, i.e., ".. that if the cause had not been, the effect never had existed." In this formulation, the *counterfactual dependence* of effects on causes is defined as $A \, \square \rightarrow C$; in words, if A were true, then C would also be true. Notationally, $\square \rightarrow$ means $A \, \square \rightarrow C$ is true in world W iff

(a) either there are no possible A-worlds (i.e., there exists no world in which A is true).
(b) or some A-world in which C holds is closer to W than any A-world in which C does not hold.

Interestingly, this statement differentiates between causal dependence and actual causation. Causal dependence is defined as

$$(O(c) \, \square \rightarrow O(e)) \wedge (O(c) \, \square \rightarrow \neg O(e))$$

where c and e are causally dependent events and $O(x)$ denotes the proposition that holds only in those worlds in which event x occurs. Actual causation in this study is transitive whereas causal dependency need not be. An important problem with this account is, as pointed out by Shoham [207], the difficulty of finding a similarity measure between possible worlds.

Another view, posited by Taylor [213], claims that causation is only a philosophical category. It can be instrumental in solving other problems or used in the analysis of other relationships but no *other* concepts can be used to analyze it. In particular, he attacks the notions of temporal causality and necessity and sufficiency. In this metaphysical view, causes and effects can be contemporaneous. Causation is not considered to have anything to do with temporal ordering of events nor with necessity/sufficiency of the cause for the effect. It is defined as the act of an agent which produces a change of state.

Domotor sums up the philosophical studies of causation in [47]. After expounding the logical, probabilistic and analytic viewpoints, he describes the desirable properties of a descriptive theory in set-theoretic terms. He states that no matter what the ontological framework of events is, the following conditions must hold:

1. $\neg A \mapsto A$ Irreflexivity — an event cannot cause itself.
2. $A \mapsto B \Rightarrow \neg (B \mapsto A)$ Asymmetry — an event cannot cause another event that is its cause.
3. $A \mapsto B \wedge B \mapsto C \Rightarrow A \mapsto C$ Transitivity — if an event causes another event that in turn is the cause of a third event, then the first event is the cause of also the third one.
4. $A(s) \mapsto B(t) \Rightarrow s < t$ Time coherence or precedence — the cause precedes the effect in time. Also, cause and effect must be *spatio-temporally adjacent*, that is they should be physically "near" each other and both should occur within a "small" interval of time.
5. $0 < P(A) < 1, \ 0 < P(B) < 1$ Both the cause and the effect have a non-zero and non-one probability of occurrence.
6. $P(B|A) > P(B|\neg A) > 0$ The cause and the effect have a positive and asymmetric conditional probability of co-occurrence; they are statistically correlated.

In the above,

" $\mapsto$ " is the causal operator;

A(x), the assumed cause and B(x), the assumed effect are events occurring at time x;

P(A) is the probability of event A occurring, and $P(A) = 1- P(\neg A)$;

$s < t$ means time point s precedes time point t;

$\neg$, $=>$, $\wedge$ are logical negation, implication and conjunction, respectively.

Domotor identifies two fundamental issues a causal study should explain:

(a) What are the events, called causes and effects, which are supposed to be in a causal relationship?

(b) What kind of relation or operator is associated with or attributed to the causal entities?

Scientific inquiries attempt to find a common invariant that can answer the first question, and a general method of causal inference to answer the second one.

The *cognitive map* by Nakamura, Iwai and Sawargi [161] is a causal network constructed from documents or surveys of past events. The nodes contain "concepts" without any detailed dynamic or physical description. The causal links between them can have one of the following eight values:

Relation	Description	Set Value
A $\overset{+}{\longrightarrow}$ B	*A* promotes *B*	$\{+\}$
A $\overset{-}{\longrightarrow}$ B	*A* retards *B*	$\{-\}$
A $\overset{0}{\longrightarrow}$ B	*A* has no effect on *B*	$\{0\}$
A $\overset{\oplus}{\longrightarrow}$ B	*A* does not retard *B*	$\{0,+\}$
A $\overset{\varnothing}{\longrightarrow}$ B	*A* does not promote *B*	$\{0,-\}$
A $\overset{m}{\longrightarrow}$ B	*A* has any effect on *B*	$\{+,-\}$
A $\overset{u}{\longrightarrow}$ B	All three relations can exist (universal)	$\{+,-,0\}$
A $\overset{a}{\longrightarrow}$ B	Ambivalent (conflicting assertions about this link have been made)	$\{\}$

One can find the net effect of one concept on another in a series-parallel network of causal links by simplifying the network. Serial chains can be simplified by 'multiplying' constituent values of the links according to the rules

$$x * y = y \quad \text{if } x = + \text{ and } y \in R$$
$$x * y = 0 \quad \text{if } x = 0 \text{ and } y \in R$$
$$x * y = a \quad \text{if } x = a \text{ and } y \in R - \{0\}$$
$$x * y = + \quad \text{if } x = - \text{ and } y = -$$

(Here $R = \{+, -, o, \oplus, \emptyset, m, u, a\}$)

Parallel chains can similarly be simplified by 'addition' of the values of the constituent links according to the following rules:

$$x \mid y = y \quad \text{if } x = 0 \text{ and } y \in R$$
$$x \mid y = a \quad \text{if } x = a \text{ and } y \in R$$
$$x \mid y = x \quad \text{if } x = y \in \{+, -\}$$
$$x \mid y = u \quad \text{if } x = + \text{ and } y = -$$

The skeleton maps, in the form of specialized semantic networks, provide the overall causal structure, and causal graphs contain the results of the causal analysis. Finally, it should be noted that such systems are not guaranteed not to contain inconsistent information — different and conflicting results may be arrived at along different pathways in the network.

Several systems have been established with the view of identifying causal relations, such as Burks' new modality of *causal necessity* within modal logic [28]. He used the logical implication with a "necessary" quantifier for the causal relation. Hayes developed a similar system, the *logic of actions* [99]. Cox and Pietrzykowski [34] use first-order logic to find the cause of some event expressed as a wff, Θ. Let the knowledge base K be a set of wwfs, such that $K \Rightarrow \Theta$ does not hold. The conjunct C is called the cause of Θ iff $(C \wedge K) \Rightarrow \Theta$ holds. The algorithm to compute C is based on linear resolution (on the clausal form of K with $\neg\Theta$) and reverse skolemization. Unfortunately, this algorithm is of combinatorial complexity (related to the cardinality of K) and it does not guarantee that the causes found are minimal.

7.1.3. The Sociological View of Causality

Causal studies in social sciences have focused on two main areas: (a) the analysis of goal-directed behavior and causal attribution, and (b) the building of statistical causal models based on consensus/population data.

MacIver attempts to classify the modes of *why*-questions according to the kinds of explanation sought [141. There is the "non-causal *why*" which does not ask for an explanation but looks for the values and standards of behavior. For instance, "you should do this because it is customary and proper" gives a reason which is not a cause. The "causal *why*" asks for a connection of dependence or interdependence between a phenomenon and an observed regularity. The causal *why* can be further categorized into physical, psychological, and social *why*s. The latter two belong to the causality of a conscious being. They have to do with *objective, motivation, design* and *social conjuncture*, as explained below.

The difference between an objective and a motivation is defined as the former having an intended end-result and an explicit set of actions to pursue the objective in question, whereas the latter does not have these. The *why* of design is moot in regard to its causal nature. It is a pattern of behavior socially imposed or culturally accepted, such as "why did you build a house of this style?" Social conjuncture is the result of group actions and it does not necessarily have a purpose. The result is not foreseen by the masses but comes out of collective behavior. Examples are "why are ground values high in the centers of cities?" In this sense, social conjuncture is analogical to the statistical analysis of a phenomenon.

Shaver's theory of "blame" [202] is a causal attribution theory based on three premises: (a) the human agency, meaning that human beings are the causes of changes in state; (b) the necessity of agency, meaning that humans perceive cause-effect relation as a necessary relation, not just constant conjunction emerging from the past; and (c) the precedence of cause over effect in time. Legal reasoning was studied in [10,11].

S. Duval and V. H. Duval's theory of causal attribution emphasize cognitive aspects rather than regularities [50]. They rely on goal-directed models of cognition in which an action is seen as a process of achieving a *preferred* or *goal state*. Causal attribution is defined as the tendency of consciousness to establish (or re-establish) the preferred state.

A statistical causal model in social sciences is, in simple terms, a set of recursive equations on observed variables, as stated by Blalock [14]. For instance, one may have the equations

$$X_1 = e_1$$
$$X_2 = c_{21}X_1 + e_2$$
$$...$$
$$X_{i+1} = C_{i+1,i}X_i + C_{i+1,i-1}X_{i-1} + .. + e_{i+1}$$
$$...$$
$$X_n = C_{n,n-1}X_{n-1} + ... + e_n$$

as a causal model where the X_is are the variables, the C_{ij}s are the correlation coefficients, and e_is are the error terms. It is assumed that $X_1, X_2,.... X_i$ precede X_{i+1}, for all i. Thus there are $(n-1)n/2$ potential causal connections for a system with n variables. The goal is to solve the set of n equations for the regression coefficients C_{ij}. The variables are from population data, measured in units of deviation from the mean in order to eliminate constant terms. As a concrete example in Blalock's four-variable system, the variables are defined as (a) division of labor, (b) residence patterns, (c) land tenure and (c) system of descent [14].

Testing a model involves eliminating some unknowns and getting multiple solutions. (Recall, only a system with n equations and n unknowns provides a unique solution.) If the resulting equations are not mutually consistent, the solution is discarded. Eliminating an unknown by setting, say, $c_{ij}=0$ means ruling out a causal connection between X_i and X_j.

The problems with statistical models in general have been pointed out before. The additional concerns about statistical causal models include the need for simplifying assumptions, having to resort to error terms when the observed variables do not converge to a pattern, the subjective judgment used on similarity between the responses to questions, and the task of how to cope with multiple causation (see, e.g., [15]).

7.1.4. *The Psychological View of Causality*

The fundamental goal of causal reasoning, whether by man or machines, is to find out what causes what and why. Depending on the form of input data and the nature of the domain, the precision and the direction of causal inferences change. The only invariant is that the cognitive process of causal understanding is strictly *inductive*. In other words, a causal inference is always made in light of what is known at the time of analysis. It evolves as more knowledge is acquired; earlier hypotheses can be invalidated, augmented or altered to give a causal account of the relation between new evidence *and* the information from which the old hypotheses were derived.

Another facet of human causal understanding has to do with commonsense. People have the tendency of deriving causal explanations by simply observing the conjoining of events in several circumstances and expect to see the same conjunction in the future. This "jumping to conclusions" phenomenon is part of our everyday lives and is at times promoted to the term "inductive inference."

The reasons for doing things are also perceived as causes of things to occur. Thus, a causal connection is made from the goals and plans of agents to their consequences. The causal associations do not come out of single, isolated incidents of using goals and plans. Rather, the history of their usage and their results constitute the pieces of memory that is called "commonsense."

The commonsense understanding of causality cannot be easily dismissed as being subjective and prone to errors and misjudgment. It is a fundamental part of human reasoning and perception. It is also an indicator that the human perception of causation is very closely tied to the memory organization and recall. The storage and retrieval of past experiences and concepts dictate the way the new information is absorbed.

Studies in Cognitive Psychology mainly consist of goal-directed and non-goal-directed models. In goal-directed systems, consciousness is viewed as an attempt to satisfy a particular state. In contrast, memory-based models emphasize the role of attention, memory and recall in causal attribution.

Piaget [175] identifies 17 categories in developing children's perception of causality. These range from rather primitive types, such as "magic", to more advanced ones resembling logical deduction. Interestingly, Piaget considers what he calls "phenomenistic causality", derived from a simple co-occurrence of events in space and time, among the most primitive types of causal perception. Abelson and Lalljee [1] conceive causal explanations as the problem of connecting the thing to be explained with some available, previously experienced pattern to fit the circumstances. The successful performance of an action sequence involves an intentional chain which begins with a goal and links with the execution of a plan for the goal [211]. The failure or success of the goal invokes pre-established (prototypical) patterns of explanation [182]. The prototype theory of psychology has also been used in Artificial Intelligence models, such as in Schank's scripts [198].

The relation between the success of recall from memory and causal relatedness of events is discussed by Trabasso and van den Broek [215], and Trabasso and Sperry [216]. Their finding suggests that the importance of a set of events in text depends directly on the direct causal relation between events and their membership in a causal chain. This also suggests an episodic memory organization of coherent events being causally related to each other. These studies have later been confirmed by Myers, Shinjo and Duffy [160].

A study of dual memory versus single memory models for events is presented by Whittlesea [223]. Dual memory suggests an independent formation of episodic and abstract (semantic) information. A single memory, however, is driven by the encoding of events as experienced and the abstracting of information from episodes — thus creating local semantic memories for episodes. This latter view has been shown to be more plausible than a dual memory explanation.

7.1.5. *Causality in Linguistics and in Natural Language Understanding Systems*

The term *causative form/verb* in linguistics refers to a *causer theme* and a *caused theme* as well as an assumed counterfactual dependence of the latter on the former [148, 204]. Causative forms include verbs such as *force, cause, make, kill,* etc. (Note that *die* is non-causative whereas *kill* is.) "I forced John to go" expresses a causal relationship because the state change is initiated by the subject, and the change would not have taken place had the subject not acted. In contrast, "I regret that John went" does not express a causal relationship since "John went" is not dependent on the subject's action. Other transitive verbs which involve state changes also signal causal connections. Shibatani, in the referenced work, also elaborates on the semantics of several types of causation, which varies depending on the verb and the type of noun-phrase (e.g., animate) used for the subject and the object of the sentence.

Other studies on syntax of causative forms include Aissen's [2], Au's [6] and Goodall's [90]. Aissen, for example, characterizes the syntax for the causal construction in general as shown in Fig. 7.1.1. Here the causal association signalled by the verb is between the first noun phrase (subject) and the object. He also shows the universality of causal constructions by demonstrating that only the order of the verb and the object changes in the above structure with languages having a lexical order Subject-Object-Verb.

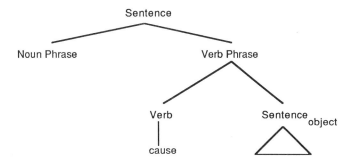

Fig. 7.1.1 — The syntax for causal construction in a sentence

Vendler [218] lists several verbs that are the results of a transformation from a non-causative form to a causative one. These include the *raise-rise, fell-fall, set-sit, lay-lie* pairs. Vendler questions the view that, in causative form, the agent of the act is attributed with the cause. He states that the agent is not an event hence cannot be in the causal chain. This apparent lack of ontology of causes and effects has plagued many causal studies in the literature.

In Schank's Conceptual Dependency (CD) theory [197], natural language utterances are converted into primitive acts, such as the physical or mental transfer of objects, which are independent of the actual language used. The CD primitives are able to differentiate the causal roles in the sentences: *result causation* is an event causing another one, *enable causation* is an event allowing another event to happen, *initiate causation* is a mental state which makes people think about events, and *reason causation* is an act following that awareness [199]. The CD representation is the main ingredient of the system of event scenarios, called *scripts.* A script is a chain of events describing a typical situation. Script-based reasoning, such as in Cullingford's SAM [38], is able to fill in the causal gaps in the scenario with a simple causal chain by assuming that if an anticipated event in a script has occurred, the ones preceding that event in the script must have also occurred. This kind of reasoning works only if the cause of each event can be attributed to a single event. Higher-order reasoning with CD forms and scripts are shown in Wilensky's program PAM [224] in which the actor's intention of performing actions (plans) guide the system to understand the input. The causal structure underlying scripts has also been applied by Read to the theory of causal attribution in social sciences [182].

The causal relations between agents' plans and their actions have been studied by Litman and Allen [139]. The purpose of this model is to give an account of utterance-plan relationships. Meta-level plans, called *discourse plans*, can be distinguished from the actual plans pursued. The interplay of discourse and topic plans enable the model to understand subdialogs in naturally occurring conversations (e.g., asking directions). Other relevant works include [17, 18, 51, 94, 108, 118, 130, 189, 226, 227].

7.1.6. *Causality in Artificial Intelligence*

Our everyday goals and plans are formed in making use of our knowledge about probable causal and temporal relations between events. Humans have developed cognitive control structures which enable us, on the basis of our past experience, to expect unseen events from recognized events, to search for a certain event in a relatively small set of related ones, to match causally connected events and assess the strength of their causal association. We rely in this activity on certain context-specific meta-knowledge, such as quasi-numerical statistical information, laws of nature, indirectly acquired ("second hand") data.

To solve complex tasks in AI, programs have to acquire and make explicit and active use of causal knowledge. Human actions, real or simulated, are usually not logically but causally determined. On the basis of a known causal relation, we may suspect a cause for a given effect. This type of reasoning, *causal analysis* or *retrodiction,* is often used in expert systems. Or we may imagine that a certain event has taken place and then assert its causal consequences. This type of hypothetical reasoning is useful for *thought experimentation* and *contingency planning*. Accordingly, there can be two different interpretations of the statement "*A* causes *B*":

- If *A* has happened, *B* will possibly happen as an effect of *A*;
- If *B* has happened, *A* might have happened as a cause of *B*.

The first relation, *progressive causality*, is a sufficient condition and means that the occurrence of the cause *A* is a sufficient premise for the occurrence of *B* to be possible. The second relation, *regressive causality*, is a necessary condition and means that the non-occurrence of *B* is sufficient for the non-occurrence of *A*. The former type of causality is useful, for example, in predicting the behavior of a system in some qualitative simulation work whereas the latter one would assist in identifying the cause from a given effect.

Unlike the logical and analytical models, most of the computational studies are actually *causal models* of some phenomena in a chosen domain and not *models of causation* itself.

Temporal causal reasoning was studied by Findler and Chen [65]. Point and duration events are distinguished; the latter may have well-determined or fuzzy *starting time*, *duration* and *finishing time*. Further, temporal relations between two duration events can be classified as one event having taken place *fully* or *partially before*, *during* or *after* the other. One point event can happen *before*, *simultaneously with*, *at about the same time as*, or *after* the other one. Reasoning about possible causality is based on first-order logic and temporal axioms about *transitivity, irreflexivity,* the existence of *predecessors* and *successors*, and the notion of *precedence* of cause before effect in time. In addition, the user can interactively set various logical, temporal and relational *constraints* that must be satisfied by the causally related events to be retrieved.

Rieger and Grinberg [188] developed a unique system in which they implemented link constructs for causal connections as programming primitives. The distinction between one-shot and continuous causal influences is also useful but the system cannot handle recognition problems.

Pople's INTERNIST [177] and CADUCEUS [178] analyze medical decision making strategies based on differential diagnosis and the underlying causal graph for diseases. CASNET, a causal-associational network by Weiss, Kulikowski and Amarel [222], represents a causal model of glaucoma in which diagnostic decisions are made by using pseudo-probability values standing for degrees of causal connections between causes and effects. In fact, the system uses the causal model as a guide for its inference engine. Patil, Szolovits and Schwartz [168] formulated a medical diagnostic system ABEL in which multiple levels of descriptions exist for medical hypotheses — from phenomenological associations on the surface to deep causal ones.

De Kleer and Brown [45] define a qualitative model of natural systems in which causes and effects are derived from qualitative differential equations (*confluences*). The recent increase of interest in *qualitative* or *naive physics* (see, e.g., the work by Bobrow [16, Forbus [82], Kuiper [117] and others) is also relevant to our topic. The common concern is to know how the behavior of a system in time — the components and the connections between components — is derived from its structure, in view of the propagation of causal effects through the connections.

Allen and Hayes [3] founded axiomatically their first-order temporal relations — *after, before, overlap*, etc. — on the single relation of contiguity between two events, *meet*. The system can reason about temporal events in a commonsense manner. Shoham [206] combined temporal logic, non-monotonic logic and the modal logic of necessity, leading to what he calls the *logic of chronological ignorance*, to deal with changes in a formal system. The modal operator **K** in the formula **K**x means "x is known". Causal theories are derived in the form

$$\Phi \wedge \Theta \mapsto K(t_i, p_i)$$

where Φ is a conjunction of sentences of the form $K(t_j, p_j)$, Θ is a conjunction of sentences of the form $\neg K(t_j, p_j)$, $t_i < t_j$ for all i and j. In this formalism, the causes (conjuncts of Φ and Θ) are disabling rules that can prevent the effect from happening. Time is discrete and the propositions are point events.

In *probabilistic causal models*, the term "probable reasoning" is used instead of "probabilistic reasoning" by some researchers (see, e.g., Shafer's inspiring work [201]) to emphasize the non-random characteristics of linking causes to effects under uncertainty. Most of the studies in this area employ subjective probabilities in the form of Bayesian inference (e.g., the work by Kim and Pearl [109], and Peng and Reggia [173] or the Dempster-Shafer theory of evidence based on belief functions by Liu [140], Ginsberg [87], Hummel and Landy [101]). In fact, the Dempster-Shafer theory is a generalization of Bayesian inference. The degree of belief in a hypothesis can vary from certainty to "don't know", and as the evidence is gathered, the system gradually reduces the number of possible hypotheses.

In *Bayesian systems*, the nodes in a causal graph are variables and the edges denote the Bayesian conditional probabilities. A priori distributions of probabilities are either assigned arbitrarily or estimated from previous knowledge. Collected evidence (conditional probabilities) are used by Glymour, Scheines, Spirtes and Kelly [88] along with the a priori distributions to come up with the a posteriori distributions of probabilities. The recognized problems with this approach are the need for large amounts of evidence, the concern about the assumption of independence between the causes and the handling of partial evidence.

Kim and Pearl [109] present a Bayesian approach to diagnostic reasoning. Causal knowledge is viewed as a hierarchical network of Bayesian probabilities, rather than a tree, which can incorporate multiple causes. (The tree structure can be maintained by aggregating the causal factors into a single variable.) Another method for diagnostic reasoning is given by Peng and Reggia [173]. This scheme tries to combine symbolic causal knowledge with Bayesian probabilities. It is shown that the use of symbolic knowledge (in the form of structural information about associated entities) can reduce the number of hypotheses generated by the model and relax the independence assumption.

Liu [140] presents an evidential reasoning system for expert systems, in which the antecedents and consequents of the rules are categorized by their causal roles. In this system, production rules are seen as causal associations among the premises and the conclusions. The causal roles may be *associational, supportive, adverse, sufficient, necessary,* or *contrary.* These roles are associated with the edges of the causal network along with the degrees of belief in the strength or weakness of the causal connection. The structure of causal knowledge in expert systems has also been studied by Gabrielian and Stickney [87]. Petri-nets are proposed as knowledge structures to handle time and probabilities in dynamic domains.

Zadeh [229, 230, 231] introduced a form of syllogistic reasoning where fuzzy quantifiers act as conditional probabilities. Six different syllogisms provide an inference system that can handle imprecise or unreliable premises to derive conclusions with certainty factors. Kosko [112] views causality as a fuzzy relation on causal concepts and defines a causal algebra to traverse a *fuzzy cognitive map* to find causal relationships. Fusaoka and Takahashi [85] present a method for representing and reasoning about progressive causation and regressive causation, defined earlier, using modal logic. The model is based on a tree structure of possible worlds. The reasoning method is based on a tableaux procedure leading to a causal model. They have also illustrated the use of the method in qualitative and hypothetical reasoning about closed domains.

The default logic of Pearl [172] is used for commonsense reasoning. Default rules are classified as either *expectation-evoking* or *explanation-*

evoking. The former leads to evidential support (association) and the latter to causal support. An implicit censorship of causal rules over evidential rules enables the system to decide that, no matter how strong the evidence may be, the effect cannot take place if it is censored (overridden).

Finally, we discuss briefly some relevant aspects of *machine learning* since it has a lot to do with inductive and causal reasoning. Considering the dynamic nature of perceiving and representing causal knowledge, the importance of learning and abstracting causal concepts should be obvious. Nevertheless, none of the studies mentioned so far in this section has any genuine learning power. The only limited learning capability they possess is *learning by being told.* That is, the models do not really learn but are imposed either the causal relationships between events or the structural and functional properties of the objects and events pertaining to finding causal associations. The studies mentioned in the following differ from this philosophy widely for they emphasize the *learning component* of causal reasoning.

DeJong's learning technique called *explanatory schema acquisition* is an attempt to learn causal relations [44]. A schema in this system corresponds to a generalized version of a concept. For instance, a schema for kidnapping incidents have the role of specific participants which are then replaced by variables (for example, the $100,000 ransom converted to "ransom money", the name John replaced by the concept "kidnapper", etc.). Mooney and DeJong base the schema on some of the CD primitives as well as on predicates about other schemata, and represent as a case-frame structure [157]. The system works by first understanding the input, and then finding whether it fits into an already acquired schema or a new schema needs to be generated. Several criteria are used to eliminate spurious generalizations, such as the novelty and generality of the goal pursued. When causal links are established in the new schema, it can be used in understanding other phenomena. Lebowitz [126] used a similar technique for generalizing from natural language text using given stereotypical information.

Lebowitz [127] also proposed a hybrid scheme which combines *similarity-based* and *explanation-based* learning. The former requires an examination of several instances of a concept to find regularity. Explanation-based learning, on the other hand, is a knowledge-intensive learning paradigm. There is a need for structural and functional information about objects to infer causal relationships. This can be done with a single example if there is sufficient information available in it.

The program by Pazzani, Michael and Flowers, called OCCAM [170, 171] tries to combine correlational information with prior causal theories (experience) for learning. For domains of which the system has an extensive knowledge, it prefers the prior causal theories over correlational information, which help the system figure out the relevant parts of the input usable in generalizations. In unfamiliar domains, the postulations about causality and intentionality help find the causal relationships but do not lead to generalizations. The generalizations are in the form of *if-then* rules relating objects and actions [170]. This model of learning needs either correlational information or prior causal theories about the domain.

Anderson [5] questions the use of strong domain theories in the simulation-based and explanation-based models of learning. Since the regularities in the domain are embedded in these theories, learning loses its inductive character. He proposes to use a causal induction method based on *contiguity* in time and space, *similarity* of causes and effects, and *statistical* evidence. The latter refers to information on how many times effect occurred when cause had taken place and, in turn, when the effect seldom occurred in the absence of the cause. The learning scheme is based entirely on the above three inductive rules. A working model is developed to detect some common mistakes students make while learning introductory algebra.

Another learning scheme by Lewis [137] to augment the similarity-based and the explanation-based models of learning is called *analysis-based* learning. In this model, a set of attribution heuristics take over the generalization process when prior causal knowledge is lacking. A particular heuristic, called *loose-end heuristic*, can be used when an action cannot be connected to a goal, and the system knows that one component to be explained is not accounted for. The failure is then attributed to the unexplained user response. The domains suitable for this paradigm tend to be well-known or well-studied areas — such as teaching algebra or programming languages — in which the validity of inputs and responses can be evaluated and checked fairly precisely.

It should finally be pointed out that causality has a ubiquitous role in human cognitive activity. It can be found in human dialogs; contributes to commonsense, qualitative and hypothetical reasoning processes; forms the basis of thought experiments (concerning, for example, a tree-like structure of possible worlds, each representing alternate courses of events), contingency planning, trouble shooting, accident investigations, medical diagnostics, etc.

7.2. The Causal Modelling System CMS

7.2.1. Research Objectives

We have wanted to establish a decision support system which can use a causal model of a given task domain as its knowledge base. Task domains are to be selected from the numerically-oriented world of science and technology. The causal model should be created and, if necessary, modified interactively by the user. The model is in the form of a directed graph in which the nodes hold the information about relevant events and the arcs represent causal relations.

Our goal was to be able to fuse knowledge from several sources (see Fig. 7.2.1.), such as

- the results of observation of the environment statistically corroborated (empirical derivation),
- information gleaned from human experts (guided learning and commonsense reasoning),
- codified laws of nature (rules of inference and analytical derivation).

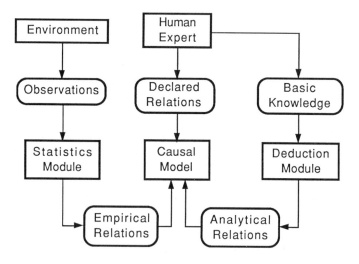

Fig. 7.2.1. — Knowledge sources for the causal modelling system CMS

CMS should differ from other work in this problem area in five important ways:

- *Event instances* should be *parametrizable* (spatial and temporal specification). The user should be able to define also general *event types*, rather than having to deal with all possible instances. Common event classes can be then stored on library files for later use. The user can also easily deal with temporally recurring events (e.g., causal biological cycles). Causal relations, defined in terms of participating events, also become parametrizable, which fact allows the expression of *classes of causal relations* (cf., laws of nature) and significantly increases the power of the system.
- The system should store in its knowledge base, in addition to the causal graph, declarative information about the *objects* and their *attributes* of the task domain. This enables the user to define an event not only in terms of other events but also in terms of objects, their properties and relation to other objects. The attributes can assume numerical or symbolic values, allowing both *quantitative* and *qualitative reasoning*.
- Events and causal relations should have *probabilistic properties* to characterize *levels of confidence*. This can express how sure the system is that an event instance has actually occurred or how strong a particular causal link is.
- A *tree of causal graphs* is to be built in the course of the graph construction process. Namely, when the system recognizes a possible inconsistency or when it performs disjunctive actions, it will fork the line of construction and build a graph for each decision choice in parallel.
- The above feature could lead to a combinatorial explosion (the world is *very* uncertain). The user has to guide the construction of the knowledge base in an interactive manner. Iterative cycles of construction and analysis is to be followed until the user is satisfied with the results. Such compiling the information from different sources (user declarations, definitions made by other users, library files of common definitions and laws of nature, assumed causal relations statistically corroborated) creates a static model that will increase the efficiency of graph traversal used in the final, question answering phase. The causal model should also be displayed in an informative way.

Summing up, we wanted to build a knowledge base constituting a causal model of a given task environment, in which the causal hypotheses inductively generated would be corroborated, modified or rejected as more information becomes available. The system should be able to handle inexact, incomplete, probabilistic[1] and fuzzy information by assigning or computing confidence factors to empirically derived or human-taught causal relations. The user can ask a variety of questions and the answers given by the system are the result of an automatic analysis of the knowledge base.

7.2.2. *The Development of the Causal Model*

The simplest form of symbolical representation of a causal relation is

$$\text{Event A} \underset{T,P}{\longmapsto} \text{Event B}$$

where T represents the temporal properties (e.g., the propagation delay between cause and effect) and P stands for the probabilistic properties of the relation. This can be generalized to

$$\text{Boolean-Event A} \underset{T,P}{\longmapsto} \text{Boolean-Event B}$$

where a *Boolean-Event* is an arbitrary Boolean expression of events (for example, the negation of the event on the right hand side indicates the retardation of the effect), which provides the full power of the different causal links postulated in [161] discussed before.

We define an *event* as either an atomic entity (to serve qualitative models), or a state or change in the state of the environment. A state is a set of values associated with the relevant objects in the task environment. (This applies, of course, to our discrete model. In the "real" universe, there is a collection of *continuous processes* in which a particular event is a *process segment* truncated by an appropriate time-window.) An implication of the above is that the system must model also individual objects and must keep track of its associated values — as a function of time and on which branch of the tree of causal models the attribute in question assumes that associated value.

---

[1]Often, the term *possibilistic* is used to refer to domains in which there is not enough satisfactory quantitative information about probabilities and, therefore, the mathematical machinery of probability theory cannot be employed.

The event can thus be considered a *conjunctively joined set of constraints* on object attribute values, which can be symbolic or numerical. For symbolic attributes, the allowed constraint can be expressed as the relation

ATTRIBUTE (OBJECT) *EQOP* VALUE

where *EQOP* $\in$ {=, $\neq$}. For numerical attributes, the corresponding relation is

ATTRIBUTE (OBJECT) *RELOP* VALUE

where RELOP $\in$ {=, $\neq$, <, >, $\leq$, $\geq$}. When one of the above relations appears on the left hand, cause side, it is interpreted as a *condition*. On the right hand, effect side, it becomes an *assignment operation* to the attribute of a single value, a set of values or a continuous range — whichever the case may be.

Several decisions had to be made concerning computations. Values associated with a numerical attribute may be obtained by performing arithmetic operations possibly also on *ranges* of values. The returned value in such case is the largest possible range of contiguous values. Further, *event durations* present a problem. Out of several possible approaches, we have chosen the following one. An event has a specific *start time* after which it keeps on occurring until its defining conditions are violated by another event's action or a specific primitive procedure returns a *terminating duration time* as part of the condition of a causal relation. ("Things remain the same unless changed by some agent".)

The underlying design philosophy has been that unless the system knows what to do in a certain situation (i.e., it has the complete information necessary), it does not do anything. This is manifested, for example, in the evaluation of *causal relation conditions*. Since the user is not required to specify all the initial attribute values for every object during the causal graph construction, it is possible that the system will attempt to evaluate an expression involving undefined values. When this happens, the condition of the causal relation fails and its effect is not considered.

The next issue to discuss concerns *inconsistent effects*. If there are no values possible that can satisfy all constraints, no action takes place until the *forking of possible environments* is initiated. For each constraint, a separate environment is generated and the appropriate action is computed. There are several problems with forking, the most serious of which is the danger of *combinatorial explosion*. Suppose we have a condition with *N* disjuncts, with the i-th disjunct having C_i conjuncts (assuming a normalized disjunctive-conjunctive Boolean expression of unique Boolean variables). We could then have $2^N - 1$ environments forked in which the condition is satisfied, and there are

$$\sum_{i=1}^{N} (2^{C_i} - 1)$$

environments in which the condition fails. *User control* (and self-constraint) is only a partial answer to this problem.

Objects are organized into an *inheritance hierarchy* of a tree structure. A special object "universal" is at its root, representing a superclass of all objects in the tree. Each node stands for the class of objects below it. *Inherited attribute values* are constants for the entire class of objects or can be used as default values if a specific value needed is not available. Such a mechanism also enables the user to define parametrizable *event classes* and *causal relation classes* in terms of *object classes*.

There are four kinds of *object attributes* declared:

· constant,
· numerical (dimensioned in English or SI units),
· object class (taking on other objects as values),
· enumerated (assuming symbolic values from a list of possible ones).

An *event instance* describes a state or a change in the state of specific object attributes and occurs at a specific time — in other words, they are not spatially or temporally parametrizable. Each forms an actual node in the causal graph, produced during the *graph construction phase*. Its defining expression consists of a conjunctively joined set of object attribute constraints occurring during a specific time interval.

Causal relation instances describe causal relations between sets of event instances, thus they are not parametrizable either. They form the arcs in the causal graph produced in the graph construction phase, and are defined in terms of the Boolean-event instances for the cause and the effect.

An *event type* is a class of event instances which can be spatially and temporally parametrized. Its definition is a Boolean expression whose atomic elements are either (possibly parametrized) object attribute constraints, event instances, or (possibly parametrized) event types.

A *causal relation type* is a class of causal relation instances which can also be spatially and temporally parametrized. Its defining cause and effect expressions are of the same form as those of event types.

An *environmental constraint* enables the user to truncate the development of impossible, unlikely or uninteresting environments. Its form is the same as that of an event type and thus can be parametrized by typed variables.

A *confidence factor* is a real number between 0 and 1. It serves as a *relative measure of quality* and not as objective probability. CMS provides mechanisms for propagating such factors through the causal graph. The *meaning* and the *role* of a confidence factor vary with which entity it is applied for. If the entity is

- an *event* or *causal relation instance*, it expresses the level of user's confidence about the validity of its definition or the level of the system's confidence that a given entity should be linked into the causal graph;
- an *event type*, it expresses the level of user's confidence that the named event will occur, given that its defining conditions are true;
- a *causal relation type*, it expresses the level of user's confidence that the named effects will occur, given that the named causes occur. (Thus it is, in fact, a subjective conditional probability value.)

The *value* of the confidence factor for an *effect* is computed as the product of two confidence factors — one in the cause expression and the other is C, the confidence associated with the causal relation type,

$$\text{Cause} \overset{\longmapsto}{c} \text{Effect}$$

as it is created and linked into the causal graph. Similarly, the value of the confidence factor for the other entities is as follows. For

- an *event instance*, it is simply the associated confidence factor;
- an *event type*, it is the product of the confidence factors of the event type and of the event type's defining expression. The latter is the minimum confidence value of the object attribute values referenced in the expression whereas the confidence of an object attribute value is the confidence of the *last* event instance constraining it;
- a *Boolean expressions*, it is the confidence of the atomic elements maximized over ORs and minimized over ANDs (similarly to the algebra of fuzzy sets).

Changes occur in the environment as a consequence of modifying object attribute values. This latter action may have three kinds of effect:

- the new information overrides any conflicting previous information;
- the new information represents a change in the world,
- the new information represents additional data about the current environment and is to be used in conjunction with the previous information.

These cases are handled in the CMS by the use of *priority factors* attached to the causal relation instances and types. When a potential conflict arises between two causal relations, the relation with the higher priority factor overrides the other one. With equal levels of priority factors, the system tries to constrain the environment so that both relations stay valid (e.g., in case of numerical ranges). If this is unsuccessful, the causal relation associated with the later time point takes precedence.

The *representation of time points* is in terms of *uncertainty intervals* defined as being between its minimum and maximum values possible.

As discussed before, the graph construction phase produces a *tree of causal graphs*. The root represents the initial environment, and the siblings of a node are actually parallel graph constructions due to inconsistent assumptions (see Fig. 7.2.2). A root-to-leaf pathway defines a consistent causal graph. Some of these may be marked as *unsuccessfully terminated* — due to a required event instance not being linked into the graph or to an overriding environmental constraint having been applied. Such causal graphs are of no more interest. The remaining pathways form valid causal graphs, which will be analyzed during the graph traversal phase. Each causal graph is an *AND/OR graph* (see Fig. 7.2.3) in which the nodes are event instances and the arcs represent conjunctive (*A* AND *B* cause *C*) or disjunctive (*A* causes *C* and *B* causes *C*) causal relation instances.

7.2.3. *The Operation of CMS*

There are three phases of operation with CMS.

(1)   In the *Declaration Phase,* the user can

- define objects, event types, event instances, causal relation types, causal relation instances, and environmental constraints;

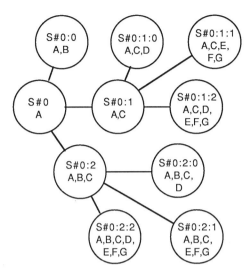

To be declared:
A occurs initially,
A causes B or C,
C causes D or E,
E causes F and G.

Fig. 7.2.2. — The tree of causal graphs

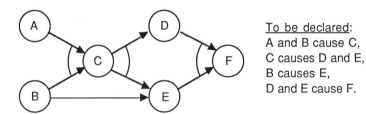

To be declared:
A and B cause C,
C causes D and E,
B causes E,
D and E cause F.

Fig. 7.2.3. — A causal graph

- merge different files of definitions to form the current environment;
- display, modify and delete existing definitions;
- save declarations to files and retrieve them from files.

(2) In the *Graph Construction Phase*, the system can

- construct a new graph using current definitions,

- save a graph to a given file and retrieve a graph from a given file for further analysis.

When constructing a graph, the user is first *prompted* for

- start and stop times of the process and, with it, the temporal interval to be analyzed, and whether the graph construction goes forward or backward in time;
- confidence thresholds to be used for pruning off unwanted worlds from the causal graph tree;
- initial values of any object attributes, which initializes the whole environment.

The graph construction is an *event-driven* cycle, each successive step of which is determined by the new time interval that contains a *window of co-occurring events*. As graph construction proceeds in a depth-first manner, new environments may be forked off as described before. A pathway is terminated when a user-specified time limit is reached or when no more change in the environment is foreseen or due to some constraint imposed. There are two types of windows of co-occurring event instances:

(a) the *current window* whose time interval contains all currently co-occurring event instances; and

(b) the *forward window* whose time interval contains all event instances that are scheduled to occur in the future.

(3) In the *Graph Traversal Phase,* the user first selects a causal graph for analysis on the basis of one of the following conditions:

- A simple enumeration yields the next environment;
- The environment with the highest average confidence is selected;
- The environment selected contains a given event instance of interest with the highest confidence.

There are seven different *analyses* that are possible to perform on the causal graph selected:

- The graph can be *displayed*, similarly to a Gantt chart, with vertical lines connecting causes and effects. There are different symbol strings for indicating causal propagation time, uncertainty in start and stop times, times of an event's occurrence, causes, effects, terminated events, and events causing the termination of other events. The graph can also be stored in a file or printed onto 80-column wide pages.

- Events in a causal graph can be displayed also in a *narrative form*, complete with causes and effects for each. If so desired, the corresponding file can also be printed;
- The user may wish to find the *common causes* of a given set of events, if any. The system identifies the cause of each event on the user-provided list and displays the intersection of the cause sets.
- Similarly, the user may wish to find the *common effects* of a given set of events, if any. The system identifies the effect of each event on the user-provided list and displays the intersection of the effect sets.
- The system can return the *principal* (with the highest confidence) or *all causes* of a given event and display it in a narrative form.
- The system can also return the *principal* (with the highest confidence) or *all effects* of a given event and display it in a narrative form.
- The user can ask the system whether two events are *causally related*. The system first finds all causes and effects of one event. A non-null result is returned if the other event is a member of this set.

Another important query would be to determine what events could have *prevented* a certain final situation, such as how could an accident have been *avoided*. Most questions of this type fall into a more general category in which two or more Boolean combinations of events are given as *states of the environment*, and the system is asked to find all causal relations connecting the events. This operation can also take on the form of an extrapolation forward or backward in time from or to a pre-defined state.

Finally, one should note here that we can freely *extrapolate* causal chains since, assumedly, every effect must have a cause and, in turn, must be the cause of something else. (We are not "concerned" with the creation and the destruction of the universe.) There can, of course, also be certain "dangling events" at the edge of the temporal horizon. The CMS can take care of those by adjusting the time window appropriately, so that the maximum terminating time of current events serves as the limiting time point for the window.

7.2.4. *Issues of Implementation*

Similarly to the systems described in previous chapters, CMS is highly interactive. The user is given the option of choosing items from a hierarchy of different menus. He is prompted for specific information when needed and the answers are checked for feasibility and coherence. The graphics display helps in visualizing causal connections over time but it is not as user-friendly as those offered with some commercially available systems of various kinds. An important feature is that before the system forks off an environment, it describes to the user the reason for it and asks him whether he would like to investigate that branch.

To show the power and versatility of the menu system, the top two levels of the hierarchy and a few lower levels of options are shown below. At the bottom level of the hierarchy, the user is guided through the information gathering phase by prompted questions.

1.   Declaration Phase Menu
1.1. Objects
1.2. Event Types
1.3. Causal Relation Types
1.4. Event Instances
1.5. Causal Relation Instances
1.6. Environmental Constraints
1.7. Include/Write File Declarations
1.8. Initialize Declarations
1.9. Display Current Definitions
R.   Return to Previous Menu

— CMS Main Menu —
1. Declaration Phase
2. Graph Construction Phase
3. Graph Traversal Phase
Q. Quit

2.   Graph Construction Phase Menu
2.1. Construct Graph
2.2. Save Graph to a File
2.3. Retrieve File from a File
R.   Return to Previous Menu

3.   Graph Traversal Phase Menu
3.1. Display Graph Statistics
3.2. Enumerate Valid Worlds
3.3. Analyze a World
R.   Return to Previous Menu

1.7.   Include/Write File Declarations Menu
1.7.1. Save Current Declarations to a File
1.7.2. Include Declarations from a File
R.   Return to Previous File

1.8.   Initialize Declarations Menu
1.8.1. Objects
1.8.2. Event Types
1.8.3. Causal Relation Types
1.8.4. Event Instances
1.8.5. Causal Relation Instances
1.8.6. Environmental Constraints
1.8.7. Include/Write File Declarations
1.8.8. Initialize Declarations
1.8.9. Display Current Definitions
R.   Return to Previous Menu

1.9.   Display Current Definitions Menu
1.9.1. Enumerate Objects
1.9.2. Enumerate Event Types
1.9.3. Enumerate Causal Relation Types
1.9.4. Enumerate Event Instances
1.9.5. Enumerate Causal Relation Instances
1.9.6. Enumerate Environmental Constraints
1.9.7. Display All Definitions
R.   Return to Previous Menu

3.3.   Analyze a World Menu
3.3.1. Select from Enumeration
3.3.2. Select by Highest Average Confidence
3.3.3. Select by Highest Event Confidence
R.   Return to Previous Menu

3.3.3.   Select by Highest Event Confidence Menu
3.3.3.1. Display Graph
3.3.3.2. Display Legend
3.3.3.3. Find the Principal Causes of an Event
3.3.3.4. Find the Principal Effects of an Event
3.3.3.5. Find the Common Causes of a Set of Events
3.3.3.6. Find the Common Effects of a Set of Events
3.3.3.7. Determine if Two Events Are Causally Related
R.   Return to Previous Menu

7.2.5. *Applications of CMS*

We have first run on CMS two test cases of hypothetical accident scenarios. These were of very different nature to prove the extent of applicability of the system.

In the first test case, several events have led to the final effect, a hypothetical *aircraft accident*. At every stage of the causal chain, a number of factors have contributed to an intermediate effect which, in turn, has become one of the contributing intermediate causes for the next stage. The events modelled were all non-cyclic, discrete events. The simplified story is as follows. The pilot spilled his coffee which then shorted out a low-fuel-level indicator light. This by itself was not enough reason for the accident because there are also other fuel gauges in the cockpit. However, the pilot did not realize the problem about the first indicator light and failed to check the others. This fact caused one tank of the plane to run out of fuel. By the time the pilot could change fuel tanks and restart the engines, it was impossible to avoid the accident.

During graph construction, forkings have taken place due to effect disjunctions. The final accident was initially declared as a non-required event instance with zero confidence level. This way, the environment in which the accident was "most likely" to occur would have the accident event with the highest confidence — whereas in the environments in which the accident did not occur, that event would retain its zero level confidence.

The environment supporting the occurrence of the accident with the highest confidence was correctly selected out of the four environments completed during graph construction. The causal chains leading to the accident were found and displayed.

The second test case was an *industrial accident* scenario — an enclosed tank was heated to the point of exploding. Here, the temperature regulator has failed and the technician, who would normally be able to discover and correct such a condition, was found to be sleeping due to several reasons. This example has shown how information from multiple sources can be fused together into a single graph representation. The information sources were (a) a natural law (the perfect gas law), (b) a relation found statistically (the behavior of employees at the plant in question), and (c) specific data about the accident. The process modelled was the heating of gas in a container of constant volume.

It is also interesting to see how causal relation types may be used to specify processes. Since they are not constrained to a particular time interval, a given set of causal relation types can form a cycle of recurring events. The four causal relation types used were (a) the action of the heating element, (b) the perfect gas law, and finally (c) and (d) represented the "imperfections" in the process; namely temperature and pressure dissipation to the environment. Although no additional worlds were forked off in this example, the tank was shown to heat up over time, eventually reach a critical pressure and explode.

7.3. The Causal Modelling System NEXUS

7.3.1. *Introduction and Research Objectives*

NEXUS views the basis of causal reasoning as pragmatic knowledge which can be learned from natural language texts. In order to account for the dynamic acquisition of this type of knowledge, a memory-based reasoning method and a learning facility are necessary. The reasoning process combines commonsense knowledge with qualitative laws of nature, such as those postulated by the spatio-temporal conditions of causal relations. The learning component acquires new knowledge and changes the interpretation of prior knowledge without human guidance.

The methodology of this ongoing research effort has its roots in the philosophical, social, psychological and linguistic views of causation outlined earlier. The notion of spatio-temporal causality and regularity theory (constant conjunction) have been obtained from philosophy. Social sciences have lent us the model of goal-directed behavior, and the relevance of goals and plans in the actions of agents. Psychology has taught us the role of memory in the detection of causal relations between past and new experiences. From linguistics, we have borrowed the structure of events and the semantics of causal utterances for the stages of parsing and semantic analysis. These linguistic concepts have proven useful in extracting intra-sentence causal connections.

All the above facets of causation have been integrated in a framework of memory-based reasoning to detect and to learn causal relations as part of *world knowledge*. It should be pointed out here that this study makes no claims about solving the outstanding problems of the concept of causality. Our objective is to find some common ground among several perspectives on causality, and to demonstrate the computational feasibility of abstracting causal knowledge within the framework adopted.

Our working hypothesis is that causal relations can automatically be learned from natural language texts using a memory-based causal reasoning process. Practically all natural language processing systems show only a peripheral interest in identifying causality in text. We have wanted to extend these models by providing

- a domain-independent, unguided, inductive learning scheme;
- a memory-based causal reasoning technique which relies on episodic memory structures and a set of causal heuristics aimed at the discovery of causal relations;
- a heterarchical commonsense reasoning scheme in which the goals of the actors/agents mentioned in text are detected by a top-down program component, whereas the plans which are used to pursue the goals (and their causal consequences) are extracted from the text using a bottom-up program component;
- a knowledge representation scheme that would enhance the derivation and manipulation of causal hypotheses in a unified framework.

The domain-independent program NEXUS is to implement these ideas as a working model. The intended use of NEXUS is to understand causal connections in "soft" domains of political and sociological nature, in which no well-defined domain theory is available. The NEXUS model is an attempt to observe how commonsense notions of causality can be integrated with more formal approaches to establish causal knowledge over a broad spectrum of human behavior. There are no a priori causal theories embedded in the learning and reasoning schemas. NEXUS will merge the episodes in its memory as causal connections and similarities between them are identified.

7.3.2. *The Causal Reasoning and Learning Schemas*

As stated above, NEXUS is designed for the task of commonsense understanding of causation from text. In order to accomplish that, it processes coherent sets of events called *episodes*, one at a time, and relates them to what it has analyzed before. Its output consists of

- a causal graph which can be used to explain the current episode — the graph is based on what is given about the episode and what NEXUS has learned from similar episodes;
- a long-term memory structure augmented by the new episode and its associations with past episodes.

During this process, learning is accomplished using certain *principles of causality*. These principles — the basic premises of the learning model and also the source of causal heuristics to be described later — subsume the desirable conditions of a descriptive theory of causality, noted in Section 7.1.2, and are as follows.

(1) Causation is *irreflexive*. That is, an event cannot cause itself. In procedural terms, since a given event can be expressed in different ways in natural language, the model needs to discover the identity of several events using event attributes and their values.

(2) Causation is *asymmetric*. Two events cannot cause each other. This is reflected in causal heuristics by making the assertion that if there is a potential causal connection between two events in the direction from *A* to *B*, there can be no causal connection assumed from *B* to *A* under the given circumstances.

(3) Causation is *transitive*. One can construct causal chains along the occurrence of events by observing pairwise causal relations.

(4) Causality implies *temporal ordering*. Event *A* has to happen before event *B* in order for *A* to be the potential cause of *B*. Temporal ordering may be implicit in two causally related events which otherwise make no reference to time. (Note that the reverse of the implication is not true.)

(5) Causality implies *spatial connectivity*. This is a rather analytic notion — one that would require a domain theory — and is not used extensively by the program. It means in other words that two spatially disparate events cannot be in a causal relationship. For convenience, only the *inaccessibility* relation is defined, which assumes that any two locations — for the cause and the effect, respectively — that are not inaccessible are accessible.

(6) Causation is *goal-directed*. In particular, one can set up a reverse causal chain, starting from the goals toward the plans for achieving the goals and along the actions which result from the plans.

(7) Causality in sentences can be signalled by *linguistic clues*, such as causative verbs.

(8) A certain *regularity* between event classes that satisfy the above causal principles implies potential causal connections between the event classes.

The above rules represent the essence of memory-based reasoning and learning schemas. Event classes aggregated from past experiences are matched against the event instances in the new input to identify new causal connections. The major task originating from the concepts presented here is to accommodate the diverse notions of causality within a single framework. The following sections describe the design and the methodology adopted.

7.3.3. *On Knowledge Representation*

Concerning the ontology of events in NEXUS, an *episode* in NEXUS's memory consists of a set of events. We define an *event* again as a change of state of the participating objects through an action. Events can be classified into two main categories: *physical* and *mental events*. Physical events alter the state of objects via physical action (e.g., being hit by a car); mental events do similarly by mental action (e.g., John read a book). Physical events can cause mental events and vice versa.

An event or a state of objects can be a *cause* in NEXUS. In other words, a cause is either an event or the *agent of an event*. The state of the agent can further be represented by a mental or physical action. This scheme can account for actions that are initiated by objects being in a certain state, and not necessarily by an explicit action. For instance, in *Mary was so angry, she hit John*, the cause of *Mary hit John* is the state of Mary being angry. *Effects* are physical or mental events.

Following Schank's terminology, we denote each event by a primitive *ACT* followed by a set of attribute-value pairs making up the event. Thus, an event is represented as a list structure composed in the form

```
(act  (attribute-1  value-1)...(attribute-n  value-n))
```

The ACTs and their attributes recognized by the system are:

act: The event itself, as described by the main verb of the sentence. In the representation scheme used by the program, the event type — mental or physical — can also be inferred from the **act**.

actor: The agent of the **act**.

object: The **object** affected by the **act**.

to: The destination of the **object** as the result of the **act**. Depending
            on the type of the **act**, it could be a **location** or another **object**
            (the recipient).

from: The origin of the **object** which again can be a **location** or another
            object (the donor).

cause: Causal consequence of the **actor** performing the **act**. It is usually
            signalled by words such as *because* and detected during the parsing
            stage.

time: **Time** of the **act**. It can assume absolute or relative values on the
            time scale as well as symbolic values, such as "past" or "future".

state: The **state** the **object** is in as the consequence of the **act** (e.g.,
            hungry, sad, tired, etc.)

location: The place at which the act takes place.

Such a taxonomy has been widely accepted and variations of it have been
part of several models [195, 199]. In addition to basic acts, NEXUS also uses
higher-level structures to account for the goals and plans of the actors. We
have represented these according to the Conceptual Dependency (CD) theory
discussed before:

```
(goal  (attribute-1  value-1)...(attribute-n  value-n)
        event/plan-list)

(plan  (attribute-1  value-1)...(attribute-n  value-n)
        event/plan-list)
```

The attributes for goals and plans are:

name: The **name** of the plan or goal.
planner: The actor who has the goal or who pursues the plan.
objective: The expected outcome of the goal or plan — the reason for which
            the actor is pursuing the goal/plan.

The *event/plan-list* is a list of events or plans. The list shows the
sequence of events that need to be carried out or satisfied, in order to achieve

the goal. The values of attributes for *basic events* have the following structure:

(slot-type value)

where *slot-type* is the type of the value. A general inheritance network is established among slot types so that two values could match even though they are of different but compatible types. For instance *(driver (Michael))* matches *(human (name (Michael)))* since drivers are human beings (i.e., that the driver IS-A human can be inferred from the inheritance network).

As said before, events in NEXUS are represented in Conceptual Dependency (CD) formalism. CD theory is based on the (brave) premise that any event that can be expressed in a natural language can also be expressed in terms of a finite set of primitives called Conceptual Primitives [198]. These primitives are categorized into physical and mental actions. CD theory has been the main ingredient of many language understanding systems, including the landmark AI programs SAM [38], PAM [224], ELI [189], and FRUMP [43]. It has proven to be a good representation system in certain contexts in which the reasoning and learning processes are action-centered. Its main drawback is the lack of a sound theory for objects (nouns) [102].

Let us next consider the following excerpts from two episodes of two different but *possibly related* stories in discussing the *memory structure* of NEXUS:

(1) "John tried to bribe the officer. The officer put him in jail."

(2) "A private contractor is arrested for bribing a city official."

The first excerpt can be interpreted as: "John had the goal of getting out of some legal trouble and he chose to pursue a plan of bribing the officer". The potential causes of his being in jail can be (1a) what he had done before he offered the bribe, or (1b) for bribing the officer. In analyzing the second episode, the system would try to identify causal connections with the first episode (after finding a match between John and the private contractor, the officer and the city official, and being arrested and being put in jail). The potential cause (1a) is ruled out since it violates the constant conjunction rule (Hume's third condition for causality, discussed in Section 7.1.1) — that is, the story does not say that the private contractor has done anything illegal before offering the bribe. The program then assumes that the potential cause (1b) is more likely to explain the story (unless, of course, it violates other causal heuristics).

As the above example demonstrates, in order to relate past experiences to new ones and to change the existing causal view in the light of new information, the reasoning and learning schemas need to be based on the memory of events and on the objects participating in the events.

Memory-based understanding systems have recently begun to get recognition. Stanfill and Waltz in MBRtalk [210] view commonsense knowledge as undigested pieces of memory which are recalled to make inferences. Lebowitz' RESEARCHER [128], an understanding system based on natural language, also relies on the memory of detailed texts as acquired and used by the system. Episodic memory structures have been used by Riesbeck and Martin [190] to disambiguate natural language utterances. Case-based reasoning, a variation of memory-based reasoning with a strong domain model, has become quite popular due to its ability to index and recall cases in its memory. (See, e.g., Kolodner's work [111]).

The advantages of memory-based reasoning stem from its amenability to learning. Earlier AI systems used hand-coded semantic information about objects of interest. In contrast, memory-based systems can search through the memory to find out whether they have encountered a similar object or a similar event beforehand. Such open-ended approach also enhances the robustness of the system. In cases in which no related prior experience is found, the system can create a new category for the unique input, which may later be enriched as more episodes are processed. Fragility of earlier AI systems under similar circumstances has been notorious.

The memory structure of NEXUS consists of a *long-term memory* (LTM), a *short-term memory* (STM), and a *working memory* (WM). The LTM has *episodic*, *declarative* and *procedural components*. The episodic component contains the episodes processed by NEXUS. The declarative component has the world knowledge about the objects participating in the events of the episodic LTM. This is usually the kind of knowledge that is not explicitly mentioned in episodes but is taken for granted (e.g., human beings are animate). The procedural LTM stores the causal hypotheses, the causal reasoning heuristics, and a set of commonsense inference rules. These are described in later sections. The STM is a dynamic entity gradually constructed by NEXUS during processing. It contains the names of the episodes related to the one currently processed. It is of limited size and life span, and is refreshed every time a new episode starts being evaluated. The WM is used as a scratch pad for manipulating and matching patterns of currently examined episodes. It is similar to an episodic memory, in as much as it contains the list of events mentioned in the current episode. It too has a limited life span and is wiped out as soon as processing is finished on the current episode. The memory structure of NEXUS is depicted in Fig. 7.3.1.

Memory is an active entity in NEXUS. Its constituents drive the reasoning and learning processes — in contrast to rule-based systems in which rules dictate the reasoning, and memory (knowledge-base) is used only for look-up.

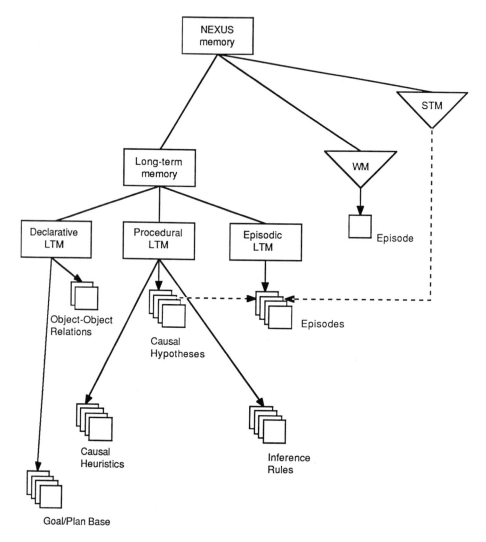

Fig. 7.3.1. — The memory structure of NEXUS

The above memory structure has been inspired by work in cognitive psychology, such as by Sanford [194]. Whittlesea's account [223] of dominant episodic memories, referenced in Section 7.1.4, has also been used extensively by our program. In Whittlesea's model, each episode has its own small set of abstract information, rather than postulating a global dual-memory in which episodes and abstract information are stored and retrieved independently of each other and in parallel.

NEXUS uses two kinds of causal knowledge: causal hypotheses and causal heuristics. The next section explains how this information is used and discusses the structure of its representation.

Causal hypotheses are derived and revised by the program as it learns about the connections between the episodes. Some knowledge representation models, such as scripts [198] and frames [155], have incorporated causal knowledge in the form of simple chains. In other words, each event in the chain causes the next event. To accommodate more complex causal connections between events, NEXUS stores causal hypotheses in the form of an AND/OR graph. Thus, the program is able to give a causal account of a single event causing several other events, as well as a single event having multiple causes. A situation of the multiple-cause-multiple-effect type is handled as a special case of multiple-events causing several single events. Figure 7.3.2 shows an example of a causal AND/OR graph.

A context-free grammar for causal hypotheses is presented below. NEXUS works only on a subset of all hypotheses that can be generated by that grammar. It can handle an arbitrary combination of causes and effects by representing them in a disjunctive form of events (see the event-list production in the grammar).

Causal-Graph	$\rightarrow$	({Node})
Node	$\rightarrow$	(**cause** Antecedent Consequent)
Antecedent	$\rightarrow$	(**agent** Event-list)
Consequent	$\rightarrow$	(**effect** Event-list)
Event-list	$\rightarrow$	(**or** {Conjunction})
Conjunction	$\rightarrow$	(**and** {Event (Means)}+)
Means	$\rightarrow$	(**means** By)
Event	$\rightarrow$	CD-event | **nil**
By	$\rightarrow$	Event | Goal | Plan

In the above, bold face characters are terminal symbols. The CD-event, Goal, and Plan are events, goals, and plans in Conceptual Dependency representation, respectively.

Causal heuristics are represented as procedures, each having three parts: *precondition*, *applicability*, and *procedure body*. These procedures are similar to the functioning of the *operators* in Fikes' STRIPS system [56]. The input to each causal heuristic consist of two events, say *x* and *y*, the current episode under evaluation, and the list of episodes related to the current episode. The output of each heuristic is a statement whether *x* can cause *y* under the given context.

Precondition is a function which determines whether the input events *x* and *y* have the attributes necessary to evaluate the heuristic (e.g., for spatial causality, the two events need to have the appropriate location attribute). Applicability of the heuristic determines whether the heuristic is applicable in the current context. If both precondition and applicability are satisfied, the procedure body returns the causal inference prescribed by the heuristic. The nature of the causal inferences are discussed in the next section.

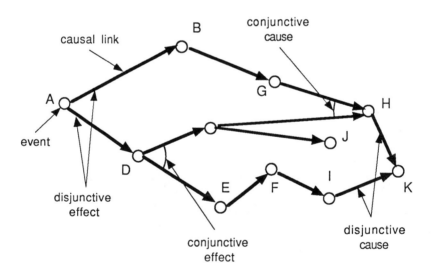

Fig. 7.3.2. — A causal AND/OR graph

7.3.4. *Reasoning in NEXUS*

We discuss now *commonsense reasoning* about goals and plans. In order to understand the actions of human agents, NEXUS detects the goals and plans pursued by them in various stages. During the parsing of the natural language input, certain states or changes in states signal the potential pursuit of a goal. For instance, when "John was hungry" is encountered, the system expects the subject John to set the goal of satisfying his hunger. Some goal states can thus be inferred by the program in a top-down manner. Nevertheless, the fact whether the subject has actually pursued the goal cannot be known until the rest of the text is processed. Hence, the system has to be aware of the ways possible to pursue that goal and has to be able to detect them if they come true. In other words, the program will have to justify the pursuit of goals, inferred in a top-down manner, by a bottom-up process. Moreover, the top-down detection of all potential goals is impossible since not all goal states can be inferred from the state changes in a single event. Such missed goals must be detected in a bottom-up manner as well.

NEXUS employs a heterarchical reasoning method for this purpose. In this scheme, the top-down component infers a set of *potential* goals while the bottom-up component detects the set of goals and plans *actually pursued*. The system acquires the knowledge about how a given goal is related to other goals and plans, and which plans can be used to pursue certain goals. The general flow of NEXUS' reasoning processes is depicted in Fig. 7.3.4.

It should be emphasized that there can yet be intentions and plans referred to in the processed text which the system has no knowledge of. It will, therefore, attempt to find causal connections using also other notions of causality, such as the concept of constant conjunction. Learning how to use a set of *given* goals is radically different and simpler than learning *unknown* goals. Automatic learning of goals, plans and their relations is one of the important future research directions which could provide insight to building self-sufficient, highly intelligent systems.

The motivation for detecting goals and plans stems from the assumption that human behavior is basically *goal-oriented*. Thus, by perceiving the reasons behind actions, one could make a causal linkage from intentions to the pursuit of goals via actions. This view of causality in goal-directed behavior has been popular in social studies of causation (see, e.g., [30] and [50]).

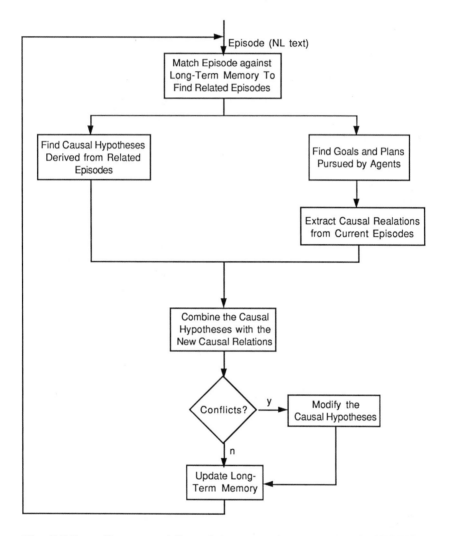

Fig. 7.3.3. — The general flow of the reasoning processes in NEXUS

Next, we discuss the problem of *causal reasoning*. The goals and plans of actors in episodes provide insight to the potential causes and effects of their actions. In addition to such goal-directed reasoning, the spatio-temporal properties of events and their co-occurrence in multiple episodes signal causal connections. To extract most likely causes and effects out of potential ones, the program must utilize a set of causal heuristics derived from the several different accounts of causation mentioned in Section 7.1..

We have explained the structure and contents of causal heuristics above. Our task is now to describe how they are used in making causal inferences. A causal heuristic may return a negative, a positive or no causal inference at all. Let $A=(x, y, E, S)$ be an ordered tuple where x and y are input events, E is a set of episodes the events of which are related to the events in the current episode S. A is provided for the analysis of each heuristic. The task of the heuristics is to find out whether x and y are causally related, given E and S. A positive causal inference is of the form

```
(cause  (agent  x)  (effect  y)  (by-means-of  event))
```

where by-means-of denote the mechanism by which x causes y. It can be inferred from the sets E or S. A negative causal inference asserts that x cannot cause y in the given context. It is of the form

```
(not-cause  (agent  x)  (effect  y))
```

A negative causal inference is more powerful than a positive one because of the falsity-preserving property of inductive hypotheses. A positive inference can be retracted or invalidated as more episodes are processed. A negative inference, on the other hand, maintains its validity once even a single piece of evidence has supported it. For instance, if event y occurs on a single occasion before event x, the only valid conclusion is that x cannot cause y. On the other hand, "y caused x" is an inductive statement which can never be completely proven as more episodes are processed (unless an "irrefutable" law of nature gives it support).

A heuristic will return no inference if its precondition, applicability or procedure body fail. An exemplary heuristic used for extracting temporal causality is given below:

```
(define-heuristic                         ;time  precedence
   '(and (basic_cd  x)  (basic_cd  y))    ;applicability
   '(and (filler_role   'time x)          ;precondition
         (filler_role   'time y))
   '(and (earlier y x)                    ;body
         (push (negative_cause  x y)      ;inference
         *causal_inferences*)))
```

We list other causal heuristics in a slightly different, space saving form below. Their role in hypothesis formation is also explained.

Heuristic	Description	Type
$(x,y) \in \mathbf{S}$ $(x \equiv y) \Rightarrow x \not\mapsto y \wedge y \not\mapsto x$	Irreflexivity. Cause and effect cannot be the same event.	Refute
$(x,y) \in \mathbf{S}$ $x \mapsto y \wedge y \mapsto z \Rightarrow x \mapsto z$	Transitivity. Used for question answering.	Instrumental
$(x,y) \in \mathbf{S}$ $x \mapsto y \Rightarrow y \not\mapsto x$	Antisymmetry. Two events cannot cause each other.	Refute
$(x,y) \in \mathbf{S}$ $time(x) \in a(x),\ time(y) \in a(y)$ $time(y) < time(x) \Rightarrow x \not\mapsto y$	Time-precedence. If y occurs before x, x cannot cause y.	Refute
$(x,y) \in \mathbf{S}$ $loc(x) \in a(x),\ loc(y) \in a(y)$ $\neg(loc(x) \xrightarrow{a} loc(y)) \Rightarrow x \not\mapsto y$	Spatial causality. If location of y is not inaccessible from location of x, x cannot cause y.	Refute
$(x,y) \in \mathbf{S}$ $\forall i,\ i \leq n,\ (x_i, y_i) \in \mathbf{E_i}$ $time(x_i) < time(y_i) \Rightarrow x \mapsto y$	Constant conjunction. If y happens every time x happens, x is the potential cause of y.	Induce
$\forall i,\ i \leq n,\ (x_i, y_i) \in \mathbf{E_i}$ $x \mapsto y \in \mathbf{H}$ $(x,y) \notin \mathbf{S} \Rightarrow x \not\mapsto y$	Negative conjunction. If y happens every time x happens, but not both appear in $\mathbf{S}$, x cannot be the cause of y.	Refute
$(x,y) \in \mathbf{S}$ $agent(x)$ has plan X; y leads to X. $x \mapsto y$ via X.	Plan-directed causality. Goal X of $agent(x)$ is the cause of y.	Induce
$(x,y) \in \mathbf{S}$ $agent(x)$ has plan X, which leads to goal y. $x \mapsto y$ via $goal(plan(X))$	Goal-based causality. Goal of $agent(x)$ is the cause of y.	Induce

The following notation has been used with the above:

$\mapsto$	:	causal operator ($x \mapsto y$ denotes x causes y),
$\not\mapsto$	:	negative causality ($x \not\mapsto y$ denotes x cannot cause y),
a(x)	:	attribute set of event x,
loc(x)	:	location attribute of event x,
time(x)	:	time attribute of event x,
agent(x)	:	agent (actor) of event x,
$\xrightarrow{a}$	:	asymmetric accessibility relation,
x, y	:	events,
H	:	the set of causal hypotheses,
E	:	the set of related episodes,
E_i	:	the i-th episode in E, $E=\{E_1, E_2,.., E_i,.., E_n\}$,
S	:	the current episode (a set of events)

$$\text{Type} \quad : \quad \begin{cases} \text{Refute — ruling out a causal connection.} \\ \text{Induce — finding a potential causal connection.} \\ \text{Instrumental — used for question answering.} \end{cases}$$

Note that, theoretically speaking, irreflexivity and antisymmetry implies time-precedence. But in practice, time-precedence deals only with those events that have time attributes whereas irreflexivity and antisymmetry reflect on the general properties of events.

7.3.5. Causal Learning in NEXUS

Causal reasoning derives tentative explanations of a phenomenon, based on the information available during the analysis. As the knowledge of the system expands, the explanations may have to change. Even though the merit of an explanation can increase if it can explain more and more data, inductive hypotheses are falsity-preserving and thus cannot be proven.

In NEXUS, positive causal inferences state and affirm potential causal relations while negative inferences rule out implausible causal relations. A causal graph is constructed in relying on plausible causal relations extracted from the current episode and on the causal hypotheses derived from related episodes. A causal graph, once constructed, is expected to answer questions on progressive and regressive causality. (As defined before, progressive causality refers to the process of finding the effects of given events, and regressive causality is the process of finding the causes, given the effects.)

In order to accomplish this, the events in a causal graph are both forward and backward chained. An exemplary forward chain in a causal graph is as follows (omitting "by-means-of")

```
(              (cause  (agent    (or  (and A)))
               (effect (or  (and B)  (and D))))
      (cause (agent    (or  (and B)))
               (effect (or  (and G))))
      (cause (agent    (or  (and C)))
               (effect (or  (and H)  (and J))))
      (cause (agent    (or  (and D)))
               (effect (or  (and C E))))
      (cause (agent    (or  (and E)))
               (effect (or  (and F))))
      (cause (agent    (or  (and F)))
               (effect (or  (and I))))
      (cause (agent    (or  (and G)))
               (effect (or  (and H))))
      (cause (agent    (or  (and H)))
               (effect (or  (and K))))
      (cause (agent    (or  (and I)))
               (effect (or  (and K))))
      (cause (agent    (or  (and J)))
               (effect (or  (and nil))))
      (cause (agent    (or  (and K)))
               (effect (or  (and nil)))))
```

Note that an agent represents always a singular event so that the system can traverse the causal graph from a given cause to several effects. Similarly, a backward chain is formed by linking the singular effects together. For instance, the node in the backward chain, corresponding to the effect H, in the above list structure is

```
(cause (agent (or (and C G)))
         (effect (or (and H))))
```

Once the backward and forward chains are completed, the system is ready to answer questions on causes and effects of a given event in the causal graph.

Learning in NEXUS involves identifying the causal links between events within the structure of causal associations. Causes and effects are classified into four basic categories, in accordance with their causal mapping: disjunctive causes, conjunctive causes, disjunctive effects, and conjunctive effects. Disjunctive causes are alternative ways of causing an event and disjunctive effects represent several distinct outcomes of an event. Disjunctions account for the differences in the causal mechanisms over several episodes. For instance,

```
(cause (agent  (or  (and A)))
       (effect (or  (and B) (and D))))
```

means that A caused B in one episode and D in another but B and D did not occur in the same episode. Conversely, the node

```
(cause (agent  (or  (and H) (and I)))
       (effect (or  (and K))))
```

denotes the fact that in one episode, K is caused by H and in another by I.

Conjunctive causes and effects signal the concurrence of events within a single episode. For example, the conjunctive cause of event H in Fig. 7.3.2 is represented as

```
(cause (agent  (or  (and C G)))
       (effect (or  (and H))))
```

and the conjunctive effect of event D in the same figure is

```
(cause (agent  (or  (and D)))
       (effect (or  (and C E))))
```

As more episodes are processed, the causal mapping within a set of events could change. Most notably, a conjunctive cause or conjunctive effect could be rendered inadequate if not all events are now conjoined in the new episode. Another case is when a causal relation formerly assumed to be disjunctive may change into a conjunctive one, supported by some overwhelming evidence about an unknown factor not mentioned in the previously analyzed episodes.

The restructuring and derivation of causal hypotheses is a continuous process. Whenever an episode is analyzed, its related hypotheses will change. If there are no related episodes in the LTM, the system will generalize from a single example and come up with potential causal relations. The ability of learning from single examples, in spite of insufficient knowledge, is due to domain-independent causal heuristics. Without them, the program would have always needed a strong domain theory or should have gone into a guided learning mode in which an expert would supply the missing information.

A very important aspect of testing the power of the learning model is the evaluation of the plausibility of causal hypotheses. Since the underlying mechanism is not deductive, a causal hypothesis could not be verified automatically. Instead, it will have to be examined by the user and rejected if not plausible. This subjective testing of plausibility is error-prone and incomplete but cannot be done by any other means. One way of reducing the errors in judgment is to feed some actual and well studied phenomena into the model and observe whether the hypotheses come close to the causal explanations that have been gleaned beforehand.

7.3.6. *Issues of Implementation*

NEXUS has been designed to be an open-ended, large system that can serve as a unified conceptual framework for studying different approaches to causal hypothesis formation concerning events of the socio-economic nature. The main modules are shown in Fig. 7.3.4.

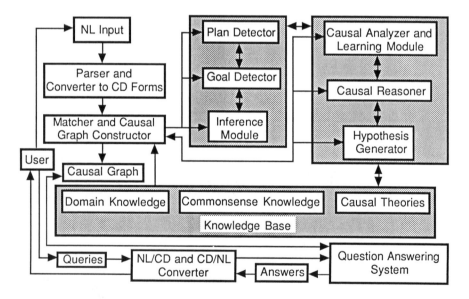

Fig. 7.3.4. — The modules of NEXUS; NL means natural language and CD is the conceptual dependency formalism

Since the implementation of NEXUS is not yet complete at the time of writing this part, the following summarizes our current view of the system operation. Text and queries are inputted in a stylized, English-like language. They are then parsed and converted into a CD representation. Episodes are identified and matched with previously stored and analyzed ones. The existing Causal Graph is augmented with the resulting segments, relying on the information obtained from the Causal Analyzer and Learning Module, the Causal Reasoner, and the Hypothesis Generator. The three constituents of the Knowledge Base — Domain Knowledge, Commonsense Knowledge and Causal Theories — are updated. Finally, a Question Answering module can respond to user queries after consulting the Causal Graph.

It is useful to establish an *inheritance hierarchy* of Causal Theories to enhance processing and to save memory. Fig. 7.3.5 shows four levels of an exemplary case.

The *hierarchy of causal theories* is helpful in the reasoning and question answering processes. Properties of and relations between entities are

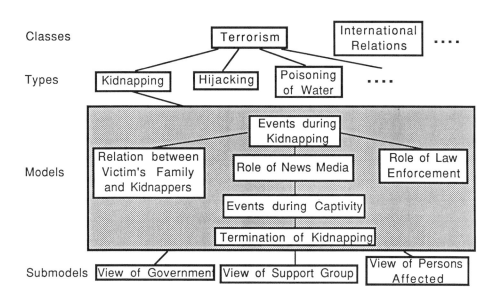

Fig. 7.3.5. — The hierarchy of causal theories

inherited downwards, and unknown values can be (temporarily) replaced by a default value from a higher level.

7.3.7. *Areas of Applications*

The applications of NEXUS will be in areas of *socio-economic* and *political* interests. The system is expected to *discover causal relations* between events from stories given in a pseudo-natural language, *answer queries* about them, develop additional *causal theories*, and *identify trends* in events and causal relations. The stories may provide no explicit information as to what event has caused what event or, even, what constitutes an event. NEXUS has to fill in the *gaps* in causal connections by using causal theories, qualitative logical-deductive methods, and probabilistic explanation techniques. It also has to *learn* in the form of inductive concept formation, and acquiring new causal theories and explanatory schemas;

In view of the qualitative nature of the domains of interest, there are serious problems concerning *data* and *result verification*, user-guided *deduction processes*, and *theory validation*. We expect to provide NEXUS with filtered data about well established stories with known causal connections and see whether the system is able to discover the correct relations by itself. We plan to carry out experiments on the effect of varying amount of prior knowledge available, user-guided learning and ambiguous data.

7.4. Summary

We have first discussed in this chapter many fundamental concepts of causation and the perspectives of several disciplines on causality. Although the problem area is of central concern in scientific discovery — in fact, in most cognitive activity — relatively little has so far been done by researchers in AI to explain and to reproduce the processes involved.

A successful system must represent and combine the essential elements of *commonsense reasoning* based on acquired *causal theories*, accumulated *general* and *domain-specific knowledge,* including qualitative statistical information and enhanced, when necessary, by certain *logical-deductive methods.*

We have also described two systems with different objectives. *CMS* aims at discovering causal relations between events in the numerically oriented world of science and technology. It is able to fuse and utilize information obtained from observing its environment (empirical derivation), interacting with human experts (guided learning), and processing laws of nature (analytical derivation). Over time, the causal hypotheses generated would get corroborated, modified or rejected as more information becomes available. The system is equipped to handle inexact, incomplete, probabilistic and fuzzy information. Empirically derived or human-taught causal relations have (computed or assigned) confidence factors associated with them.

There are several uses of the CMS: (i) to give quantitative causal interpretation to chains of past events; (ii) to fill in gaps in sequences of events; (iii) to verify dynamic process models of systems; (iv) to process events in a top-down manner (will event X occur?) or in a bottom-up manner (what event must have occurred to produce the state Y?); (v) to specify a sequence of actions to reach certain goals or to avoid certain outcomes; (vi) to evaluate certain goal-oriented behavior.

The other system described, *NEXUS* works in socio-economic and political domains. Its input information is gleaned from texts in a pseudo-natural language form, which it first converts to a conceptual dependency format. Events, plans, goals, objects and relations are analyzed and compared with its knowledge available about similar entities. A causal graph is constructed and, with more information acquired, gradually enlarged. The system uses this structure to respond to user queries about causal relations between events.

As a final comment, we wish to point out that the systems presented here could serve, after some relatively small modifications, for the basic task of *Automatic Rule Induction*. To discover rules for action is a process very important in many expert system domains. We have discussed the problem of automatic knowledge acquisition in connection with the Quasi-Optimizer in Chapter 3 and with the Advice Taker/Inquirer in Chapter 4. Causal modelling techniques could be another effective approach to it.

7.5. Acknowledgements

My thanks are due to Tim Bickmore for his work on CMS, and to Cem Bozsahin for his efforts in the development of NEXUS.

8. The Predictive Man-Machine Environment (PMME)

8.1. Introduction and Research Objectives

In the previous chapters, we have described several large-scale programming systems which aim at combining the strength and flexibility of humans and computers in decision making, problem solving and planning tasks. It seemed desirable to bring these systems together so that they all can contribute to the solution of some difficult problems. The logical scheme of this *integrated programming environment* is shown in Fig. 8.1.1.

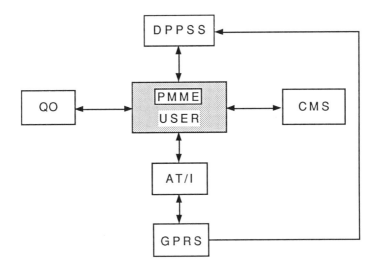

QO: The Quasi-Optimizer System,
DPPSS: The Distributed Planning and Problem Solving System,
AT/I: The Advice Taker/Inquirer System,
GPRS: The Generalized Production Rule System,
CMS: The Causal Modelling System,
PMME: The Predictive Man-Machine Environment

Fig. 8.1.1. — The logical schema of the integrated programming environment

8.1.1. *On Decision Support Systems*

We have discussed in Chapter 2 various issues in decision making and planning, and the reasons for which the analytical treatment of real-life problems is usually very difficult or impossible. Computer-based *decision support systems* often provide assistance in complex semi-structured tasks. They serve, rather than replace, human judgment, and they improve the effectiveness and efficiency of human decision making and planning by extending their range, capability and speed. There is a growing body of literature (see, e.g., [20, 107]) that describes case studies and the general principles of building decision support systems. Such systems provide flexible, robust, user-friendly and inexpensive environments and promote a reasonably high-level interaction between man and computers. Graphic displays, enhanced with color capabilities, augment human perception and make it easier for the user to assess a situation and its critical factors.

A subset or all of the following characteristics of a task environment make a decision support system for it useful and necessary:

- the database is too large for a human to access or to make even conceptual use of;
- to arrive at a solution requires significant processing of the available information;
- there is a time pressure to carry out the computations (the environment changes) and/or to obtain the solution (action is needed *on time*);
- delicate judgment is required as to which variables are relevant and what their current values are.

The above properties are valid for the problem domains we have discussed in the previous chapters. We have, therefore, decided that the programming environment integrating our systems will also serve as a decision support system.

8.1.2. *The Predictive Man-Machine System*

We have implemented, both for training and routine operations in decision making and planning, a novel type of man-machine system [62]. Figure 8.1.2 shows its essential components.

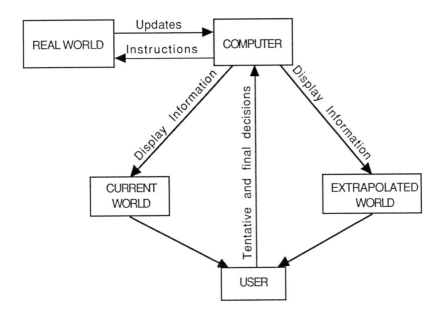

Fig. 8.1.2. — The predictive man-machine environment for decision making and planning

The *Current World* displays the essential features of the *Real World*. Events from the Real World (or, as in our case, the simulated Real World) *update* the representation of the affected objects in the computer. The updating can be controlled in three different ways:

- at regular points of time, following a user-defined frequency (*time-driven updating*),
- whenever one of the set of user-specified events takes place, an update is triggered (*event-driven updating*),
- on user's command, regardless of the time elapsed since the last update or the conditions currently satisfied by the environment (*user-driven updating*).

The user makes *tentative decisions* on the basis of the Current World displayed, say, at time t_0 and transmits them to the computer. He then specifies a time point in the future, t_1, at which he wants to see the expected consequences of the tentative decisions. This extrapolation of the status of the world is computed, on the basis of the computer-resident model, through periodic (user-specified) time increments up to the final time point t_1. However, if during the time period between t_0 and t_1, a *conflict situation* occurs, as defined by the user again, the calculation is interrupted, and the conflict situation with the corresponding time is displayed on the second unit called *Extrapolated World*. If no conflict has occurred between the time points in question, the relevant features of the world will reflect the permissible consequences of the tentative decisions on the second screen.

If the user is satisfied with the status of the resulting world, he finalizes the tentative decisions by informing the computer accordingly. Otherwise, he makes a different set of tentative decisions and goes through the above cycle as many times as necessary, time permitting.

As noted above, the image of the Extrapolated World is computed according to a computer-resident model that can, of course, be incomplete and erroneous to some extent. We have also implemented the framework of a simple learning mechanism that can increase the *predictive precision* of this model. The image of the Extrapolated World is stored and when real time "catches up" with the time point of extrapolation, t_1, the *then* Current World is compared with the stored image. The user is warned if there is a more than tolerable discrepancy between the two. The latter may indicate that either the Real World has been affected by some significant and unexpected factors during the period of extrapolation (about which nothing can be done) or that the world model in the computer is not satisfactory. In many instances (such as the one described below), a domain-specific *learning process* can cause an appropriately constructed world model to converge to a satisfactory version.

8.2.   The Simulated Air Traffic Control Environment

Each of the projects described in the previous chapters has been applied to some problem in the area of air traffic control. It seemed, therefore, logical to implement a Simulated Air Traffic Control (SATC) environment for the PMME.

8.2.1. *Some Terminology*

Some terminology needs to be established at first. To set up any simulated environment for the PMME, we have to define a set of *objects* inhabiting it. The *conditions* that affect the objects can be either constant over time (e.g., the location of a mountain range) or subject to change (e.g., current weather). Objects can also affect each other (e.g., two planes with an intersecting flight path). An object passes through different *phases* in its interaction with the environment. A phase can be considered as a distinct stage in the lifetime of the object — the period of its sojourn in the environment. During a phase, the object may perform certain *functions* toward achieving a subgoal (or, finally, a goal). This achievement of the subgoal places the object into another phase (e.g., a plane passes through eight major phases in its flight: preparation for take-off, take-off, climb, cruise, descent, approach, landing and taxiing). The overall goal is, of course, the safe and timely flight between the origin and the destination.

Associated with each object is a set of *attributes* which determine the constant and varying properties of the object — including the phase it is in. The number and the type of the attributes, in general, depend on the domain of application. The user must define the functions that can alter the values of the attributes every time a user-specified *incremental parameter* (usually time) value elapses. There is a trade-off here between "too large" and "too small" triggering value for the incremental parameter. In the former case, more computation has to be performed but there is a lesser chance that a critical conflict can get by unnoticed in between two attribute updatings. In the latter case, the opposite is true.

It is important to point out that the attributes of the objects must change independently from each other in the Current World and in the Extrapolated World. This separation of *attribute scopes* is a basic requirement when causal relations between decisions and consequences are sought.

A large knowledge base needs to be prepared for any non-trivial domain of application. For example, in case of the SATC, it must include

- the physical characteristics of one or several airports (length and location of airstrips; height and location of potential obstructions, such as mountains and towers; direction of radio beacons; etc.)
- the physical characteristics of the participating planes (symbolic notation for the manufacture/model, maximum initial landing speed, approach angle, approach speed, descent angle, descent speed, maximum and average cruising speed, climb angle, take-off speed, turning radius, fuel capacity, fuel consumption in different phases, etc.);

- dictionary and syntax needed in communicating between pilots and the air traffic controller;
- a (machine) interpreter for the above communication;
- links between controller instructions and functions which make attribute value changes accordingly.

8.2.2. *Attempts at the Implementation*

We went through three iterations before a satisfactory implementation was attained. At first, we decided to drive the displays of the Current World and the Extrapolated World from two different computer accounts which could communicate with each other through common files. This approach was slow, inefficient and error-prone.

The second attempt was based on a specially built device that could direct, under software control, graphics signals from a serial I/O port to either of the display units. The criteria of the design were simple maintainability, cost-effective and easily obtainable parts, adherence to interface standards for multi-purpose applications, and flexibility in switching characteristics and baud rates. The idea was that, upon the receipt of a special character (at a selected baud rate), the device intercepts the output stream from the host computer, toggles the output switch and transmits the data stream to the other display unit. It turned out that the device became faulty rather frequently and had to be abandoned.

Finally, in the third attempt, we decided to go along the software switching route again but, this time, using one single account for the two display units since there is no time overlap between the processing needs of the two worlds.

8.2.3. *The Operation*

Figure 8.2.1. provides a view of the SATC environment. The left-hand display unit shows the Current World and the right-hand one the Extrapolated World. The simulated radar scopes are patterned after those in use at the TRACON centers of the Federal Aviation Administration.

The displays present a map which is approximately 100 miles in radius and contain a "control section", an area for which the present group of controllers

Fig. 8.2.1. — The two graphic display units show simulated radar scopes. The left-hand one carries the Current World and the right-hand one the Extrapolated World. The controller's microphone is used to provide voice input to the simulated pilots

is responsible. The map is centered over the location of the radar dish that generates the display. The circular grid lines are shown five miles apart. The location of the airways, radio beacons, mountain peaks and other relevant landmarks are also indicated in a standard symbolic format.

In real-life, ATC displays produce only a blip on the screen for an individual aircraft while a so-called transponder on the aircraft continually sends messages to superimpose additional display information as needed. This information on the aircraft's identification, ground speed, altitude and other data is presented in a block connected with a line to the exact location of the aircraft on the screen. Our display simulates this final result.

We have kept the simulated environment as realistic as possible but also enhanced the displays with certain additional facilities that are not yet available with real-life radar scopes. We have made use of the *color capabilities* of our display units to make the identification of various groups of symbols easier. The user is also given the ability to *center the display* over any part of the map, and to *zoom* it in and out as necessary to obtain a more detailed view of potential problem areas. We have added a *system clock*, a *text window* for presenting the status of the system and the airplanes, and a *menu-display* to make the environment more powerful and flexible.

The menu system allows the user to select or change any of the simulation parameters which are not associated with the behavior of the airplanes. Menu selections are made in using a stylus and a digitizing tablet next to the display. Through this menu system, one can vary the time parameters of the simulation, change the color of a group of symbols or alter the center and scale of the map as noted above.

The next innovation concerning a *second display mode* has also been provided wherein the vertical profile of an airplane's movement can be viewed. The display in this mode is centered on the target aircraft and the viewing plane is rotated so that the target is always moving perpendicular to the controller's line of sight. Although the current ATC radar scopes do not yet have this capability either, the facility was provided to enhance the man-machine interface in certain critical situations.

The internal structure of the program handles each aircraft as an independent object, complete with its own performance and behavioral characteristics. The airplane objects carry information about their flight plan, commands previously issued to them by the controller, a full description of their current status, and a pointer to a generic block of information (an information structure of the frame type) containing the performance characteristics of a class of airplanes (e.g., DC-9, Boeing 747, etc.). This way, any environment can be described in a high-level manner, in terms of the properties of the aircraft in that area.

The data structures used allow the user to manipulate *system time* easily. To update the environment to a new time, a set of operations has to be performed on the representation of each airplane. Extrapolation is accomplished by taking the current state of the environment and then repeatedly updating it by the time increment parameter until the specified time is reached and the display is triggered. (In this, it is assumed that no conflict has occurred. Otherwise, the extrapolation process is interrupted, warning sound and light signals are given, and the conflict in question is displayed at that time point.) Further, the environments can be saved periodically to allow the controller the possibility of an *"instant replay"*.

Another novel addition to our system is a versatile *reminder system*. The controller is able to describe a variety of anticipated situations using the menu system. When the program senses that a predefined situation is about to occur, it alerts the controller with a tone and a message with reference to the triggering condition. This facility may save the controller valuable time by providing retrieval cues for previously generated plans.

Finally, while the controller can modify simulation parameters though the menu system, commands and requests are sent to the simulated airplanes though a simple *voice recognition system*. It was initially limited to 100 words at a time, representing a rather small vocabulary. We have then used an efficient downloading and uploading program between the voice recognition system and the mainframe computer, and completed an open-ended parser for blocks of words in the controller's communication to the pilots. This has enabled the use of several (somewhat overlapping) blocks ('conceptual segments') of 100 words.

8.3. An Evaluation of the Predictive Man-Machine Environment

In order to evaluate the PMME concept, a series of experiments were designed and performed. The objective of the experiments was to compare the efficiency and effectiveness of human planning and decision making in a complex, real-time problem domain — with and without the aid of the interactive extrapolation facility.

The experiments were based on the implementation of the SATC in the PMME described above. The task specified for the human subject was the (somewhat simplified) problem of *en route* ATC. The subject was first presented with a routing task in which only the real-time information was provided. The subject was then given a statistically equivalent (randomly generated) task but this time the extrapolation facility was available to him. The performance of the subject was automatically evaluated in each case according to the number of airspace conflicts (near-collisions) occurred during the session, and the efficiency with which the aircraft were routed through the subject's control sector. (Note that we deliberately introduced certain air corridor configurations and a large number of participating aircraft in order to create many potential conflicts for the benefit of the experiments.)

8.3.1. *The Experimental Set-Up*

In exactly one half of all experiments, the extrapolation facility was enabled so that the controller could look at "snapshots" of future situations. At the beginning of each session, the program randomly generated a number of straight-line air corridors across the control sector. As time was passing, a number of aircraft were also generated with randomly selected take-off times, airways, altitudes and air speeds. The subject could issue directives to the individual aircraft via the keyboard, the digitizing tablet, or "natural language" voice input to a simple spoken word understanding system. He could order them to change their altitude or heading, or to resume their initial "requested" flight plan. Additionally, in the experiments in which the extrapolator was enabled, the subject could ask for an extrapolation to some specified time point. The results of this was displayed as soon the corresponding computation was completed. The session was terminated when the last plane left the subject's control sector.

Typical parameters of an experimental setup were as follows.

Number of airways:	3;
Number of aircraft:	26;
Requested ground-speed range:	350-550 mph;
Requested altitude range:	19,000-22,000 feet;
Time increment for the Current World:	5 sec;
Time increment for the Extrapolated World:	10 sec;
Conflict criteria:	separation less then 5 miles horizontally and 1,000 feet vertically.

The following *statistics* were calculated and recorded at the end of each session:

- The number of conflicts — for each aircraft in a conflict situation at any time, incremented at the end of every internal time increment;
- The average "off-distance" — the difference in miles between the lengths of each aircraft's requested straight-line route and the sum of the distance actually travelled through the air traffic controller's sector and the distance from the point it left the sector to the plane's destination;
- The average "climb-distance" — the total distance in feet each aircraft was told to climb during its flight through the sector, plus the difference

between its requested altitude and its altitude upon leaving the sector (when positive).

8.3.2. The Results of Experiments

Twenty experiments, each lasting approximately 45 minutes, were performed — ten without extrapolation and ten with it. The results can be summarized as follows:

Measure	Without Extrapolation	With Extrapolation	Reduction
Average conflicts/ session	48,2	22,2	54%
Average off-distance/ plane	0,01424 mile	0,00797 mile	44%
Average climb- distance/ plane	654,6 feet	545,0 feet	17%

8.4. Summary

The experimental results clearly indicate that in the task environment chosen, the subjects' performance with the extrapolation facility was significantly better than without it. Part of the difference in performance can be attributed to the fact that the PMME was implemented on a time-shared system that was also taking care of other users. Therefore, the Current World environment would be run somewhat faster (in real-time) when the extrapolator was disabled, which fact makes the controller's task a little more difficult. However, the difference in speed was small enough (particularly with several other users on the system) that the results should still represent an accurate evaluation of the usefulness of the PMME.

Although the use of the extrapolation facility improved the subjects' performance, we have found that it can also be abused and lead to undesirable outcomes. For example,

- After having gained some experience in the use of the extrapolator, it is easy to become complacent about the control task when no conflicts may occur within the time interval between the current and the extrapolated time.
- If extrapolations are continually requested for short time intervals, the major part of one's attention is drawn away from the Current World to the Extrapolated World — again resulting in a possible oversight of conflicts in the immediate future;
- In "crisis" situations in which many decisions must be made in a short period of time (e.g., several conflicts are imminent), extrapolation can often be more of a hindrance than help. When facing such a confusing situation, a seemingly safe thing to do is to run the extrapolation into the near future. However, this action usually results in wasting valuable time during which effective directives should have been issued to resolve the conflicts.

In spite of the above caveats, PMME is a useful, general-purpose environment for real-time decision making and planning. It should, however, be relied upon only as an auxiliary source of information — the planner and decision maker should always focus his attention on the unfolding events in the Current World.

Summing up, the predictive man-machine environment can in general be used to improve trainees'

- analytical skills,
- long-range planning ability,
- diagnostic capabilities

so that they can

- balance objectives,
- associate methods with results,
- convert objectives, specified as the desired state of the environment, to dynamic process models,
- utilize resources to minimize risks,
- contrast costs and benefits.

Several other applications seem also possible, both for automated training and evaluation as well as routine operations. These range from investment portfolio management to command and control tasks in different conflict situations.

8.5. Acknowledgements

The major participants in the project were Tim Bickmore, Bob Cromp and Neal Mazur.

9. Overall Summary

We have discussed in this book a number of projects, each of which relates to some aspects of decision making strategies — *strategies*, for short. We have considered a strategy both as a guide to making plans and solving problems, and a mechanism that observes a part of its environment, regarded as *relevant* by the strategy, evaluates it and prescribes a certain response to it.

Decision making, in a general sense, represents a time-sequence of choices which are the responses of the strategy to the environment and to its history of development. The *planning process* usually precedes decision making. Planning contrasts and, in some sense, tries out alternative methods of attack in a coarse-grained manner. It applies a crude model to the task of accomplishing certain goals, and computes a sequence of action steps toward these goals — a set of desired situations. The goals can be positive (to attain certain objectives) or negative (to avoid certain events). The consequences of a decision may be known in advance completely, partially, or not at all, and in a deterministic or probabilistic manner.

We started with the description of a programming system, the *Quasi-Optimizer*, that can analyze and assess the value of distinct strategy components and of a whole strategy. It can construct descriptive theories, computer models, of such strategies, select their best constituents, and combine these in a quasi-optimum strategy that is at least as good as or better than any strategy analyzed. The quasi-optimum strategy is a normative theory in the statistical sense and within the framework of the input strategies. It should be noted that the discovery of rules — in fact, causal relations as described under Causal Modelling Systems — is also a subject of study in Cognitive Psychology with reference to concept attainment, sequence extrapolation, induction of mathematical relations and analogical reasoning. The applications of this system include the automatic acquisition of knowledge from different human experts, reconciling and fusing such pieces of knowledge, and computing a consistent and complete knowledge base for, for example, an expert system. We have also shown how a properly configured version of the Quasi-Optimizer can verify and validate discrete-event numerical simulation models.

The second project, the *Advice-Taker/Inquirer*, is also a domain-independent program that can be taught to construct, fine-tune and oversee the operation of an expert system.. During its learning phase, a human expert teaches the system interactively about the domain of application in a high-level manner. The program then monitors, during its operational phase, the resulting expert system and continually attempts to improve its level of performance. We have shown how this system can be used to teach strategies that control assembly line balancing and the operation of street traffic lights.

The next project, the *Generalized Production Rule System*, can provide point and functional estimates of values of hidden variables — variables that cannot be observed and measured at any arbitrary point of time and space. The process of estimation is based on certain stochastic and, assumedly, causal relations learned by the system between the known values of hidden variables and the mathematical behavior of related open variables. We have used this system to estimate the underlying reasons for and the often unknown consequences of strategy decisions.

We then discussed the objectives and the methodology of *Distributed Planning and Problem Solving* in general and four rather different domains for which we have implemented such distributed systems — a promising future air traffic control system, nationwide manufacturing control, the control of moving resources to be assigned to moving tasks, and the control of street traffic lights.

The next two projects described were aimed at *Causal Modelling*. The first one, *CMS*, discovers causal relations between events in the numerically oriented world of science and technology. It can fuse knowledge obtained from different sources — from empirical observations, narrative description of commonsense reasoning by humans, and analytically derived laws of nature. The second project, *NEXUS*, attempts to explain causality in socio-economic and political domains. Its knowledge is acquired from texts provided in a pseudo-natural form of English. The conclusions are the result of memory-based and commonsense reasoning processes.

The last project described, the *Predictive Man-Machine Environment* provides a convenient programming and user-oriented setting in which the above systems have been integrated. It can be used for the automatic training and evaluation as well as the routine activity of a control operator. Monitoring, supervising and improving behavior is a demanding and necessary component of learning, whether by humans or computers. There are two participating graphics display units one of which shows the essential features of the Current World, on the basis of which the user can make tentative decisions. He can ask the system to show the consequences of these tentative decisions computed for a future time point on the second display, the Extrapolated World. If satisfied, he can finalize his decisions; otherwise, he may go through this cycle as many times as necessary, time permitting. The system can provide a feedback to the user, indicating in which regions of the decision space he is proficient and in which he needs improvement. A Simulated Air Traffic Control Environment has been implemented in this framework — an appropriate choice in view of the fact that predictability in general is the major factor leading to safe operations. In this case, the stacks of planes waiting for landing at known altitude, well-separated air corridors and many other measures reduce the processing load and restrict the variable space for air traffic controllers, who are highly trained pattern recognizers.

Each of the programming systems above operate in a highly interactive manner. The user and the computer form a collaborating team, each contributing what he/she/it can do best. Finally, it may be in order to compare the relative strength of man and machines. *Man* can easily and often without any noticeable effort,

- assimilate new information;
- respond to needs of categorizing, sorting, assessing, identifying, and discarding information at different levels of abstraction;
- recognize and ignore irrelevant and unreliable data;
- make good use of redundant information in attaining robustness in cognitive tasks;
- find sensible compromises in situations of conflict which may be due to inconsistent data, clashing interests or newly emerging priorities;
- distinguish between more and less useful information;
- make global judgments derived inductively on the basis of disjoint sample values;
- apply general principles and techniques to specific cases in a deductive manner;
- discover and utilize partial similarities and analogies;
- construct and adhere to plans but be willing to modify them when the situation warrants it;
- improve his level of performance with experience;
- use tolerant and unobtrusive error detection and correction facilities;

- develop original and creative techniques which usually make use of previously available ones.

  In turn, *computers* can effectively and efficiently

- perform computations on numerical and symbolic information precisely, at a high speed and unaffected by fatigue, boredom, saturation or predisposition;
- be programmed in a flexible manner to emulate all the above human accomplishments to varying and increasing degrees of sophistication and success;
- be used for inexpensive experimentation on a wide spectrum of different approaches and techniques;
- overcome the shortcomings of the currently available software and hardware by undergoing an accelerating process of revolutionary improvements.

References

[1] Abelson, R. P. and M. Lalljee: Knowledge structures and causal explanation (In D.J. Hilton (Ed.): *Contemporary Science and Natural Explanation: Commonsense Conceptions of Causality*, pp. 175-203. The Harvester Press: Brighton, Great Britain; 1988).

[2] Aissen, J.: *The syntax of causative constructions* (Garland Publishing: New York; 1979).

[3] Allen, J. F. and P. J. Hayes: A common-sense theory of time (*Proceedings of the Ninth International Joint Conference on Artificial Intelligence*, pp. 528-531. Los Angeles, CA; 1985).

[4] Anderson, J. R.: *The Architecture of Cognition* (Harvard University Press: Cambridge, MA; 1983).

[5] Anderson, J. R.: Causal analysis and inductive learning (*Proceedings of the Fourth International Workshop on Machine Learning*, pp. 288-299. Irvine, CA; 1987).

[6] Au, T. K.: A verb is worth a thousand words — The causes and consequences of interpersonal events implicit in language (*Journal of Memory and Language*, **25**, pp. 104-122; 1986).

[7] Barr, A. and E. A. Feigenbaum (Eds.): *The Handbook of Artificial Intelligence*, Vol. 1. William Kaufmann: Los Altos, CA; 1981).

[8] Barstow, D.: A knowledge-based system for automatic program construction (*Proceedings of the Fifth International Joint Conference on Artificial Intelligence*, pp. 382-388. Cambridge, MA; 1977).

[9] Becker, S. and B. Selman: An overview of knowledge acquisition methods for expert systems (Tech. Rep. CSRI-184, Computer Systems Research Institute, University of Toronto; 1986).

[10] Bessonet, C. G. de and G. R. Cross: Representation of some aspects of legal causality (In C. Walter (Ed.): *Computer Power and Legal Reasoning*. West Publishing: St. Paul, MN; 1985).

[11] Bessonet, C. G. de and G. R. Cross: An artificial intelligence application in the law — CCLISP, a computer program that processes legal information (*High Technology Law Journal*, **1**, pp. 329-409; 1987).

[12] Biermann, A. W.: Approaches to automatic programming (In F. L. Alt (Ed.): *Advances in Computers*, **15**, pp. 1-63; 1976).

[13] Black, F.: A deductive question-answering system (In M. Minsky (Ed.): *Semantic Information Processing*, pp. 354-401. MIT Press, Cambridge; 1968).

[14] Blalock, H. M.: *Causal Inferences in Nonexperimental Research* (University of North Carolina Press: Chapel Hill, NC; 1964).

[15] Blalock, H. M.: *Basic Dilemmas in the Social Sciences* (Sage Publications: Beverly Hills, CA; 1984).

[16] Bobrow, G. D.: Qualitative reasoning about physical systems — An introduction (*Artificial Intelligence*, **24**, pp. 1-5; 1984).

[17] Bobrow, G. D. and A. Collins: *Representation and Understanding: Studies in Cognitive Science* (Academic Press: New York, NY; 1975).

[18] Bobrow, D. G. and T. Winograd: An overview of KRL, a knowledge representation language (*Cognitive Science*, **1**, pp. 3-46; 1977).

[19] Boddy, M. and T. Dean: Solving time-dependent planning problems (*Proceedings of the Eleventh International Joint Conference on Artificial Intelligence*, pp. 979-984. Detroit, MI; 1989).

[20] Bonczek, R. H., C. W. Holsapple and A. W. Whinston: *Foundations of Decision Support Systems* (Academic Press: New York, NY; 1981).

[21] Bradshaw, G. L., P. Langley and H. A. Simon: BACON.4 — The discovery of intrinsic properties (*Proceedings of the Third National Conference of the Canadian Society for Computational Studies of Intelligence*, pp. 19-25. Victoria, BC, Canada; 1980).

[22] Brand, M.: Preface (In M. Brand (Ed.): *The Nature of Causation* (University of Illinois Press: Chicago, IL; 1976).

[23] Brownston, L., R. Farrell, E. Kant and N. Martin: *Programming Expert Systems in OPS5* (Addison-Wesley: Reading, MA; 1985).

[24] Bruno, G., A. Elia and P. Laface: A rule-based system to schedule production (*Computer*, **19**, No. 7, pp. 32-40; 1986).

[25] Buchanan, B. and E. Feigenbaum: Dendral and Meta-Dendral — Their applications dimension (*Artificial Intelligence*, **11**, pp. 5-24; 1978).

[26] Buchanan, B., G. Sutherland and E. Feigenbaum: Heuristic DENDRAL — A program for generating explanatory hypotheses in organic chemistry (In B. Meltzer and D. Michie (Eds.): *Machine Intelligence*. 4, pp. 209-254. Edinburgh University Press: Edinburgh, Great Britain; 1969).

[27] Burbridge, J. L.: AI and capacity planning with GT and period batch control (In J. Browne (Ed.): *Knowledge-Based Production Management Systems*, pp. 247-264. North-Holland: Amsterdam, The Netherlands; 1989).

[28] Burks, A.: *Chance, Cause, Reason — An Inquiry into the Nature of Scientific Evidence* (University of Chicago Press: Chicago; 1975).

[29] Challou, D. J.: Towards a knowledge-based data structuring aid (Technical Report, Department of Computer Science, University of Illinois, October; 1984).

[30] Clark, N. K., G. M. Stephenson and D. R. Rutter: Memory for a complex discourse — The analysis and prediction of individual and group recall (*Journal of Memory and Language*, **25**, pp. 295-313; 1986).

[31] Clausewitz, K. von: *On War* (Transl. by J. J. Graham. Routledge and Kegan Paul: London, Great Britain; 1940).

[32] Cochran, W. G. and Cox, G. M.: *Experimental Designs* (Second ed., Wiley: New York, NY; 1957).

[33] Colmerauer, A., H. Kanoui, R. Pasero and P. Roussel: Un système de communication de machine-homme en français (Rapport de Recherche CRI 72-18, Groupe Intelligence Artificielle, Université Aix-Marseille II; 1973).

[34] Cox, P. T. and T. Pietrzykowski: Causes for events — Their computation and applications. (*Proceedings of the Eighth International Conference on Automated Deduction*, pp. 608-621. Oxford, Great Britain; 1986).

[35] Corkill, D. D. and V. R. Lesser: The use of meta-level control for coordination in a distributed problem solving network (*Proceedings of the*

Eighth International Joint Conference on Artificial Intelligence, pp.748-756. Karlsruhe, Federal Republic of Germany; 1983).

[36] Cromp, R. F.: The task, design and approach of the Advice Taker/Inquirer system (Tech. Rep. TR-85-014, Department of Computer Science, Arizona State University; 1985).

[37] Cromp, R. F.: The Advice Taker/Inquirer, a system for high-level acquisition of expert knowledge (*Telematics and Informatics,* **5,** pp. 297-312; 1988).

[38] Cullingford, R.: *Script Application — Computer Understanding of Newspaper Stories* (Ph.D. Thesis and Technical Report 116, Department of Computer Science, Yale University; 1978).

[39] Davis, R., B. G. Buchanan and Shortliffe, E.: Production rules as a representation for a knowledge-based consultation program (*Artificial Intelligence,* **8,** pp. 15-45; 1977).

[40] Davis, R. and D. B. Lenat: *Knowledge-Based Systems in Artificial Intelligence* (McGraw-Hill: New York, NY; 1982).

[41] Davis, R. and R. G. Smith: Negotiation as a metaphor for distributed problem solving (*Artificial Intelligence,* **20,** pp. 63-109, 1983).

[42] Dean, T. and M. Boddy: An analysis of time-dependent planning (*Proceedings of the Sixth National Conference on Artificial Intelligence,* pp. 49-54. St. Paul, MN; 1988).

[43] DeJong, G.: Skimming newspaper stories by computer (Technical Report 104, Department of Computer Science, Yale University; 1977).

[44] DeJong, G.: Acquiring schemata through understanding and generalizing plans. (*Proceedings of the Eighth International Joint Conference on Artificial Intelligence,* pp. 462-464. Karlsruhe, Federal Republic of Germany; 1983).

[45] DeKleer, J. and J. Seely Brown: A qualitative physics based on confluences (*Artificial Intelligence,* **4,** pp. 7-83; 1984).

[46] Diederich, J.: Knowledge-based knowledge elicitation (*Proceedings of the Tenth International Joint Conference on Artificial Intelligence,* pp. 201-204, Milan, Italy; 1987).

[47] Domotor, Z.: Causal models and space-time geometries (*Synthèse*, **24**, pp. 5-57; 1972).

[48] Doyle, R. J.: Constructing and refining causal explanations from an inconsistent domain theory (*Proceedings of the Fifth National Conference on Artificial Intelligence*, **1**, pp. 538-544. Philadelphia, PA; 1986).

[49] Duda, R. O., Gaschnig, J. G., and Hart, P. E.: Model design in the PROSPECTOR consultant system for mineral exploration (In D. Michie (Ed.): *Expert Systems in the Micro-Electronic Age.* Edinburgh University Press: Edinburgh, Great Britain; 1979).

[50] Duval, S. and V. H. Duval: *Consistency and Cognition — A Theory of Causal Attribution* (Lawrence Erlbaum: Hillsdale, NJ; 1983).

[51] Dyer, M. G.: *In-Depth Understanding* (The MIT Press: Cambridge, MA; 1986).

[52] Erman, L., F. Hayes-Roth, V. Lesser and D. Reddy: The HEARSAY-II speech-understanding system — Integrating knowledge to resolve uncertainty (*Computing Surveys*, **12**, pp. 213-253; 1980).

[53] Eshelman, L. and J. McDermott: MOLE — A knowledge acquisition tool that uses its head (*Proceedings of the Fifth National Conference on Artificial Intelligence*, pp. 950-955. Philadelphia, PA; 1986).

[54] Fain J., Gorlin, D., F. Hayes-Roth, S. J. Rosenschein, H. Sowizral, and Waterman, D.: *The ROSIE Language Reference Manual* (Techn. Report N-1647-ARPA, The RAND Corp., Santa Monica, CA; 1981).

[55] Feigenbaum, E. A.: The simulation of verbal learning behavior (In E. A. Feigenbaum and J. Feldman (Eds.): *Computer and Thought*, pp. 297-309. McGraw-Hill: New York, NY; 1963).

[56] Fikes, R. E. and N. J. Nilsson: STRIPS — A new approach to the application of theorem proving to problem solving (*Artificial Intelligence*, **2**, pp. 189-208; 1971).

[57] Findler, N. V.: An overview of the Quasi-Optimizer system (*Large-Scale Systems, Theory and Applications*, **5**, pp.123-130; 1983).

[58] Findler, N. V.: On a computer-based theory of strategies (*Kybernetes*, **12**, pp. 89-97; 1983).

[59] Findler, N. V.: On the automatic generation of descriptive and normative theories (Invited Paper; *Proceedings of the IEEE International Conference on Systems, Man and Cybernetics*, pp. 596-601. Bombay and New Delhi, India; 1983).

[60] Findler, N. V.: Some Artificial Intelligence contributions to air traffic control (*Proceedings of the Fourth Jerusalem Conference on Information Technology*, pp. 470-475. Jerusalem, Israel; 1984).

[61] Findler, N. V., M. S. Belofsky and T. W. Bickmore: On some issues concerning optimization and decision trees (*Proceedings of the International Conference on Mathematical Modelling in Science and Technology*, pp. 191-197. Zurich, Switzerland; 1983).

[62] Findler, N. V., T. W. Bickmore and R. F. Cromp: A general-purpose man-machine environment with special reference to air traffic control (*International Journal for Man-Machine Studies*, **23**, pp. 587-603; 1985).

[63] Findler, N. V., T. Bickmore, L. Ihrig and W.-W. Tsang: On the heuristic optimization of a certain class of decision trees (To appear in *Information Sciences*; 1990).

[64] Findler, N. V., J. E. Brown, R. Lo and H. Y. You: A module to estimate numerical values of hidden variables for expert systems (*International Journal for Man-Machine Studies*, **18**, pp. 323-335; 1983).

[65] Findler, N. V. and D. Chen: On the problems of time, retrieval of temporal relations, causality and coexistence (*International Journal of Computer and Information Sciences*, **2**, pp. 161-185; 1973).

[66] Findler, N. V. and R. F. Cromp: A heuristic approach to optimum experimental design (*Computational Statistics and Data Analysis*, **2**, pp. 167-178; 1984).

[67] Findler, N. V. and J. Gao: Dynamic hierarchical control for distributed problem solving (*Data and Knowledge Engineering*, **2**, pp. 285-301; 1987).

[68] Findler, N. V. and Q. Ge: Distributed planning and control for manufacturing operations (*Proceedings of AI'88, the Australian Joint Artificial Intelligence Conference*, pp. 439-455. Adelaide, Australia; 1988).

[69] Findler, N. V. and Q. Ge: Perceiving and planning before acting — An approach to enhance global network coherence (*International Journal of Intelligent Systems*, **4**, pp. 459-470; 1989).

[70] Findler, N. V. and J. van Leeuwen: The complexity of decision trees, the Quasi-Optimizer, and the power of heuristic rules (*Information and Control*, **40**, pp. 1-19; 1979).

[71] Findler, N. V. and R. Lo: A note on the functional estimation of values of hidden variables — An extended module for expert systems (*International Journal for Man-Machine Studies*, **18**, pp. 555-565; 1983).

[72] Findler, N. V. and R. Lo: An examination of distributed planning in the world of air traffic control (*Journal of Distributed and Parallel Processing*, **3**, pp. 411-431; 1986).

[73] Findler, N. V. and R. Lo: Empirical studies on distributed planning for air traffic control. Part I: A dynamic hierarchical structure for concurrent distributed control. Part II: The location centered cooperative planning system. Part III: Experimental results (Technical Report No. TR-88-007, Computer Science Department, Arizona State University; 1988).

[74] Findler, N. V., R. Lo and C. Lo: A pattern search technique for the optimization module of a morph-fitting package (*Mathematics and Computers in Simulation*, **29**, pp. 41-50; 1987).

[75] Findler, N. V. and J. P. Martins: On automating computer model construction — The second step toward a Quasi-Optimizer system (*Journal of Information and Optimization Sciences*, **2**, pp. 119-136; 1981).

[76] Findler, N. V. and N. Mazur: On the verification and validation of discrete-event simulation models (To appear in *Transactions on Simulation*; 1990).

[77] Findler, N. V., N. Mazur and B. McCall: A note on computing the asymptotic form of a limited sequence of decision trees (*Information Sciences*, **28**, pp. 207-231; 1983).

[78] Findler, N. V. and B. B. McCall: A conceptual framework and a heuristic program for the credit-assignment problem (*IEEE Transactions on Systems, Man and Cybernetics*, **SMC-14**, pp. 750-754; 1984).

[79] Findler, N. V. and E. Morgado: Morph-fitting — An effective technique of approximation (*Journal of Mathematics and Computers in Simulation*, pp. 417-429; 1983).

[80] Findler, N. V., G. Sicherman and S. Feuerstein: Teaching strategies to an Advice Taker/Inquirer system (In P. A. Samet (Ed.): *Proceedings of the EuroIFIP-79 Conference*, pp. 457-465. North-Holland: Amsterdam, The Netherlands; 1979).

[81] Findler, N. V., W.-W. Tsang and L. Ihrig: A pruning algorithm for finding all optimal decision graphs (*Proceedings of the International Conference on Optimization Techniques and Applications*, pp. 470-479. Singapore; 1987).

[82] Forbus, K. D.: Qualitative process theory (*Artificial Intelligence*, **24**, pp. 85-168; 1984).

[83] Forgy, C.: The OPS5 user's manual (Technical Report CMU-CS-81-135, Department of Computer Science, Carnegie-Mellon University; 1981).

[84] Fox, M. S.: *Constraint-Directed Search — A Case Study of Job Shop Scheduling* (Morgan Kaufmann: Los Altos, CA; 1987).

[85] Fusaoka, A. and K. Takahashi: On mechanical reasoning about causal relations (*Artificial Intelligence*, **1**, pp. 15-22; 1986).

[86] Gabrielian, A. and M. E. Stickney: Hierarchical representation of causal knowledge (*Proceedings of IEEE Western Conference on Expert Systems*, pp. 82-89. Anaheim, CA; 1987).

[87] Ginsberg, M. L.: Does probability have a place in non-monotonic reasoning? (*Proceedings of the Ninth International Joint Conference on Artificial Intelligence*, pp. 107-110. Los Angeles, CA; 1985).

[88] Glymour, C., R. Scheines, P. Spirtes and K. Kelly: *Discovering Causal Structure* (Academic Press: Orlando, FL; 1987).

[89] Good, I. J.: A causal calculus (*British Journal for the Philosophy of Science*, **11**, pp. 305-318; 1961).

[90] Goodall, G.: *Parallel Structures in Syntax — Coordination, Causatives and Restructuring* (Cambridge University Press: Cambridge, Great Britain; 1987).

[91] Green, C. C.: The application of theorem-proving to question-answering systems (*Proceedings of the First International Joint Conference on Artificial Intelligence*, pp. 219-237. London, Great Britain; 1969]

[92] Green, C.: A path entropy function for rooted trees (*Journal of the ACM*, **20**, pp. 378-384; 1973).

[93] Green, C.: A summary of the PSI synthesis system (*Proceedings of the Fifth International Joint Conference on Artificial Intelligence*, pp. 380-381. Cambridge, MA; 1977).

[94] Greiner, R. and D. Lenat: A representation language language (*Proceedings of the First National Conference on Artificial Intelligence*, pp. 165-169. Stanford, CA; 1980).

[95] Haas, N. and G. G. Hendrix: An approach to acquiring and applying knowledge (*Proceedings of the First National Conference for Artificial Intelligence*, pp. 235-239. Stanford, CA; 1980).

[96] Hagers, G and A. Hansson, A.: Logic modelling of cognitive reasoning (*Proceedings of the Eighth International Joint Conference on Artificial Intelligence*, pp. 81-83. Karlsruhe, Federal Republic of Germany; 1983).

[97] Hayes-Roth, F., P. Klahr and D. Mostow: Advice-taking and knowledge refinement — An iterative view of skill acquisition (In J. R. Anderson (Ed.): *Skill Acquisition and Development*. Lawrence Erlbaum: Hillsdale, NJ; 1981).

[98] Hayes-Roth, F., D. Waterman and D. Lenat (Eds.): *Building Expert Systems* (Addison-Wesley: Reading, MA; 1983).

[99] Hayes, P.: A logic of actions (In B. Meltzer and D. Michie (Eds.): *Machine intelligence 6*, pp. 495-509. American Elsevier: New York; 1971).

[100] Henle, M.: On the relation between logic and thinking (*Psychological Review*, **69**, pp. 366-378; 1962).

[101] Hummel, R.A. and M. S. Landy, M.S.: A statistical viewpoint on the theory of evidence (*IEEE Transactions on Pattern Analysis and Machine Intelligence*, **10**, pp. 235-247; 1988).

[102] Jackendoff, R.: *Semantics and Cognition* (The MIT Press: Cambridge, MA; 1988).

[103] John, P. W.: *Incomplete Block Designs* (Marcel Dekker: New York, NY; 1980).

[104] Jomini, A. H. de: *Summary of the Art of War* (Transl. by O. F. Winship and E. E. McLean. Putnam: New York, NY; 1854).

[105] Kanet, J. J. and H. H. Adelsberger: Expert systems in production scheduling (*European Journal in Operations Research*, **29**, pp. 51-59; 1987).

[106] Kant, E.: A knowledge-based approach to using efficiency estimation in program synthesis (*Proceedings of the Sixth International Joint Conference on Artificial Intelligence*, pp. 457-462. Tokyo, Japan; 1979).

[107] Keen, P. G. W. and M. S. S. Morton: *Decision Support Systems* (Addison-Wesley: Reading, MA; 1978).

[108] Keller, R. M.: Learning by re-expressing concepts for efficient recognition (*Proceedings of the Third National Conference for Artificial Intelligence*, pp. 182-186. Washington, DC; 1983).

[109] Kim, J. H. and J. Pearl: A computational model for causal and diagnostic reasoning in inference systems (*Proceedings of the Eighth International Joint Conference on Artificial Intelligence*, pp. 190-193. Karlsruhe, Federal Republic of Germany; 1983).

[110] Klinker, G., C. Boyd, S. Genetet and J. McDermott: A KNACK for knowledge acquisition (*Proceedings of the Sixth International Joint Conference on Artificial Intelligence*, pp. 488-493. Tokyo, Japan; 1979).

[111] Kolodner, J.: Organization and retrieval in a conceptual memory for events (*Proceedings of the Seventh International Joint Conference on Artificial Intelligence*, pp. 227-233. Vancouver, BC, Canada; 1981).

[112] Kosko, B.: Fuzzy cognitive maps (*International Journal of Man-Machine Studies*, **24,** pp. 65-75; 1986).

[113] Kosy, D. W. and B. P. Wise: Self-explanatory financial planning models (*Proceedings of the Fourth National Conference on Artificial Intelligence*, pp. 176-181. Austin, TX; 1984).

[114] Kowalski, R.: *Logic for Problem Solving* (American Elsevier: New York, NY; 1979).

[115] Kowalski, R.: The early years of logic programming (*Communications of ACM*, **31**, pp. 38-54; 1988).

[116] Kuhn, T.: *The Structure of Scientific Revolutions* (Second ed. University of Chicago Press: Chicago, IL; 1970).

[117] Kuipers, B.: Commonsense reasoning about causality — Deriving behavior from structure (*Artificial Intelligence*, **24**, pp. 169-203; 1984).

[118] Kunz, J. C.: Analysis of physiological behavior using a causal model based on first principles (*Proceedings of the Third National Conference on Artificial Intelligence*, pp. 225-228. Washington, DC; 1983).

[119] Laird, J. E., P. S. Rosenbloom and A. Newell: Towards chunking as a general learning mechanism. (*Proceedings of the Eighth International Joint Conference on Artificial Intelligence*, pp. 188-192. Karlsruhe, Federal Republic of Germany; 1983).

[120] Langley, P. W.: BACON — A production system that discovers empirical laws (*Proceedings of the Fifth International Joint Conference on Artificial Intelligence*, pp. 344-346. Cambridge, MA; 1977).

[121] Langley, P.: Rediscovering physics with BACON.3 (*Proceedings of the Sixth International Joint Conference on Artificial Intelligence*, pp. 505-507. Tokyo, Japan; 1979).

[122] Langley, P., G. L. Bradshaw and H. A. Simon: BACON.5 — The discovery of conservation laws (*Proceedings of the Seventh International Joint Conference on Artificial Intelligence*, pp. 121-126. Vancouver, BC, Canada; 1981).

[123] Langley, P., H. A. Simon and G. L. Bradshaw: Rediscovering chemistry with the BACON system (In R. S. Michalski, J. G. Carbonell and T. M. Mitchell (Eds.): *Machine Learning: An Artificial Intelligence Approach*, pp. 307-329. Tioga: Palo Alto, CA; 1983).

[124] Langley, P., J. M. Zytkow, G. L. Bradshaw and H. A. Simon: Three facets of scientific discovery (*Proceedings of the Eighth International Joint Conference on Artificial Intelligence*, pp. 464-468. Karlsruhe, Federal Republic of Germany; 1983).

[125] Langley, P., J. M. Zytkow, H. A. Simon and G. L. Bradshaw: The search for regularity — Four aspects of scientific discovery (In R. S. Michalski, J. G.

Carbonell and T. M. Mitchell (Eds.): *Machine Learning: An Artificial Intelligence Approach*, Vol. II, pp. 435-469. Morgan Kaufmann: Los Altos, CA; 1986).

[126] Lebowitz, M.: Generalization from natural language text (*Cognitive Science*, **7**, pp. 1-40; 1983).

[127] Lebowitz, M.: Integrated learning — Controlling explanation (*Cognitive Science*, **10**, pp. 219-240; 1986).

[128] Lebowitz, M.: The use of memory in text processing (*Communications of ACM*, **31**, pp. 1483-1502; 1988).

[129] Lee, C. N., P. Liu and M. Chiu: Dynamic causal model construction for model based reasoning (*Proceedings of IEEE Western Conference on Expert Systems*, pp. 90-95. Anaheim, CA; 1987).

[130] Lenat, D. B.: BEINGS — Knowledge as interacting experts (*Proceedings of the Fourth International Joint Conference on Artificial Intelligence*, pp. 126-133. Tbilisi, Soviet Union; 1975).

[131] Lenat, D. B.: The nature of heuristics (*Artificial Intelligence,* **19**, pp. 189-249; 1982).

[132] Lenat, D. B.: Theory formation by heuristic search — The nature of heuristics II: Background and examples (*Artificial Intelligence*, **21**, pp. 31-59; 1983).

[133] Lenat, D. B.: EURISKO — A program that learns new heuristics and domain concepts. The nature of heuristics III: Program design and results (*Artificial Intelligence*, **21**, pp. 61-98; 1983).

[134] Lenat, D. B.: The role of heuristic in learning by discovery — Three case studies (In R.S. Michalski, J. G. Carbonell and T. M. Mitchell (Eds.): *Machine Learning: An Artificial Intelligence Approach*, pp. 243-306. Tioga: Palo Alto, CA; 1983).

[135] Lenat, D. B. and J. S. Brown: Why AM and EURISKO appear to work (*Artificial Intelligence*, **23**, pp. 269-294; 1984).

[136] Lewis, D.: Causation (*Journal of Philosophy*, **70**, pp. 556-567; 1973).

[137] Lewis, C.: Why and how to learn why — Analysis-based generalization of procedures (*Cognitive Science*, **12**, pp. 211-256; 1988).

[138] Lindsay, R., B. G. Buchanan, E. A. Feigenbaum and J. Lederberg: *DENDRAL*. (McGraw-Hill: New York; 1980).

[139] Litman, D. J. and J. F. Allen: A plan recognition model for subdialogues in conversations (*Cognitive Science*, **11**, pp. 163-200; 1987).

[140] Liu, G. S.: Causal and plausible reasoning in expert systems (*Proceedings of the Sixth National Conference on Artificial Intelligence*, **1**, pp. 220-225. Philadelphia, PA; 1986).

[141] MacIver, R. M.: *Social Causation* (Harper and Row: New York, NY; 1964).

[142] Mackie, J. L.: *The Cement of Universe: A Study of Causation* (Oxford University Press: Oxford, Great Britain; 1974).

[143] Martelli, A. and U. Montanari: Optimization of decision trees through heuristically guided search (*Communications of the ACM*, **21**, pp. 1025-1031; 1978).

[144] Martin, W. and R. Fateman: The MACSYMA system (*Proceedings of the Second Symposium on Symbolic and Algebraic Manipulation*, pp. 59-75. Los Angeles, CA; 1971).

[145] Matheus, J. and L. Rendell: Constructive induction on decision trees (*Proceedings of the Eleventh International Joint Conference on Artificial Intelligence*, pp. 645-650. Detroit, MI; 1989).

[146] McCarthy, J.: The advice taker (In M. Minsky (Ed.): *Semantic Information Processing*, pp. 403-410. MIT Press: Cambridge, MA; 1968).

[147] McDermott, J.: R1 — An expert in the computer systems domain (*Proceedings of the First National Conference on Artificial Intelligence*, pp. 269-271. Stanford, CA; 1980).

[148] McDowell, J. P. and K. Dahlgren: Commonsense reasoning with verbs (*Proceedings of the Tenth International Joint Conference on Artificial Intelligence*, pp. 446-455. Milan, Italy; 1987).

[149] Melle, W. van: A domain-independent system that aids in constructing knowledge-based consultation programs (*Proceedings of the Sixth International Joint Conference on Artificial Intelligence*, pp. 923-925. Tokyo, Japan; 1979).

[150] Melle, W. van, E. Shortliffe and B. Buchanan: EMYCIN — A domain-independent system that aids in constructing knowledge-based consultation programs (*Infotech State of the Art Report 9*, No. 3; 1981).

[151] Michalski, R.: Discovering classification rules using variable-valued logic system VL1 (*Proceedings of the Third International Joint Conference on Artificial Intelligence*, pp. 162-172. Stanford, CA; 1973).

[152] R. Michalski, J. Carbonell, T. Mitchell: *Machine Learning* (Tioga: Palo Alto, CA; 1983).

[153] Michalski, R. S. and P. Winston: Variable precision logic (*Artificial Intelligence*, **28**, pp. 121-146; 1986).

[154] Minsky, M.: Steps toward artificial intelligence (In E. A. Feigenbaum and J. Feldman (Eds.): *Computers and Thought*, pp. 406-450. McGraw-Hill: New York; 1963).

[155] Minsky, M.: A framework for representing knowledge (In P. Winston (Ed.): *The Psychology of Computer Vision*. McGraw-Hill: New York, 1975).

[156] Mitchell, T. M.: Version spaces — A candidate elimination approach to rule learning (*Proceedings of the Fifth International Joint Conference on Artificial Intelligence*, pp. 305-310. Cambridge, MA; 1977).

[157] Mooney, R. and G. DeJong: Learning schemata for natural language processing (*Proceedings of the Ninth International Joint Conference on Artificial Intelligence*, pp. 683-687. Los Angeles, CA; 1985).

[158] Mostow, D. J.: *Mechanical Transformation of Task Heuristics into Operational Procedures* (Ph.D. dissertation, Department of Computer Science, Carnegie-Mellon University, Pittsburgh, PA; 1981).

[159] Mostow, D. J.: Machine transformation of advice into a heuristic search procedure (In R. S. Michalski, J. Carbonell and T. M. Mitchell (Eds.): *Machine Learning — An Artificial Intelligence Approach*. Tioga Press: Palo Alto, CA; 1983).

[160] Myers, J. L., M. Shinjo and S. A. Duffy: Degree of causal relatedness and memory (*Journal of Memory and Language*, **26**, pp. 453-465; 1987).

[161] Nakamura, K., S. Iwai and T. Sawargi: Decision support using causation knowledge base (*IEEE Proceedings on Systems, Man and Cybernetics*, **SMC-12**, pp. 765-777; 1982).

[162] Neumann, J. von and O. Morgenstern: *Theory of Games and Economic Behavior* (Third ed. Princeton University Press: Princeton, NJ; 1953).

[163] Newell, A. and H. A. Simon: *Human Problem Solving* (Prentice-Hall: Englewood Cliffs, NJ; 1972).

[164] Newell, A. and H. A. Simon: Computer Science as empirical inquiry — symbols and search (*Communications of ACM*, **19**, pp. 113-126; 1976).

[165] Nilsson, N. J.: Probabilistic logic (*Artificial Intelligence*, **28**, pp. 71-87; 1986).

[166] Noble, D.: Schema-based knowledge elicitation for planning and situation assessment aids (*IEEE Proceedings on Systems, Man and Cybernetics*, **1**, pp. 473-483; 1989).

[167] Pagall, G.: Learning DNF by decision trees (*Proceedings of the Eleventh International Joint Conference on Artificial Intelligence*, pp. 639-644. Detroit, MI; 1989).

[168] Patil, R. S., P. Szolovits and W. B. Schwartz: Modelling knowledge of the patient in acid-base and electrolyte disorders (In P. Szolovits (Ed.): *Artificial Intelligence in Medicine*, pp. 191-226. Westview Press: Boulder, Co; 1982).

[169] Payne, H. and W. Meisel: An algorithm for constructing optimal binary decision trees (*IEEE Transactions on Computers*, **C-26**, pp. 905-916; 1977).

[170] Pazzani, M. J.: Integrated learning with incorrect and incomplete theories (*Proceedings of the Fifth International Workshop on Machine Learning*, pp. 291-297. Ann Arbor, MI; 1988).

[171] Pazzani, M., D. Michael and M. Flowers: The role of prior causal theories in generalization (*Proceedings of the Fifth National Conference on Artificial Intelligence*, **1**, pp. 545-550. Philadelphia, PA; 1986).

[172] Pearl, J.: Embracing causality in formal reasoning (*Proceedings of the Sixth National Conference on Artificial Intelligence*, pp. 369-373. Seattle, WA; 1987).

[173] Peng, Y. and J. A. Reggia: A probabilistic causal model for diagnostic problem solving - Part I: Integrating causal inference with numeric probabilistic inference (*IEEE Transactions on Systems, Man and Cybernetics*, **SMC-17**, pp. 146-162; 1987).

[174] Pereira, F. C. N. and D. H. D. Warren: Definite clause grammars for language analysis — A survey of the formalism and a comparison with augmented transition networks (*Artificial Intelligence*, **13**, pp. 231-278; 1980).

[175] Piaget, J.: *The Child's Perception of Physical Causality* (Transl. by M. Gabain. Routledge and Kegan Paul: London, Great Britain; 1970).

[176] Politakis, P.: Using empirical analysis to refine expert system knowledge bases (Technical Report CBM-TR-130, Laboratory for Computer Science Research, Rutgers University, New Brunswick, NJ; 1982).

[177] Pople, H. E., Jr.: The formation of composite hypotheses in diagnostic problem solving — An exercise in synthetic reasoning (*Proceedings of the Fifth International Joint Conference on Artificial Intelligence*, pp. 1030-1037. Cambridge, MA; 1977).

[178] Pople, H. E., Jr.: Heuristic methods for imposing structure on ill-structured problems — The structuring of medical diagnostics (*Proceedings of the Symposium on Artificial Intelligence in Medicine*, pp. 119-190. Boulder, CO: 1982).

[179] Post, E.: Formal reductions of the general combinatorial problem (*American Journal of Mathematics*, **65**, pp. 197-268; 1943).

[180] Quinlan, J.. R.: Discovering rules by induction from large collections of examples (In D. Michie (Ed.): *Expert Systems in the Micro-Electronic Age*. Edinburgh: Edinburgh University Press; 1979).

[181] Quinlan, J. R.: Induction of decision trees (*Machine Learning*, **1**, pp. 81-106; 1986).

[182] Read, S. J.: Constructing causal scenarios — A knowledge structure approach to causal reasoning (*Journal of Personality and Social Psychology*, **52**, pp. 288-302; 1987).

[183] Reboh, R.: Using a matcher to make an expert consultation system behave intelligently (*Proceedings of the First National Conference for Artificial Intelligence*, pp. 231-234. Stanford,CA; 1980).

[184] Reboh, R.: Extracting useful advice from conflicting expertise (*Proceedings of the Eighth International Joint Conference on Artificial Intelligence*, pp. 145-150. Karlsruhe, Federal Republic of Germany; 1983).

[185] Reiter, R.: On reasoning by default (*Proceedings of the TINLAP-2 Conference*. University of Illinois at Urbana-Champaign; 1978).

[186] Rich, E.: *Artificial intelligence* (McGraw-Hill: New York, NY; 1983).

[187] Rich, C., H. E. Shrobe and R. C. Waters: Overview of the Programmer's Apprentice (*Proceedings of the Sixth International Joint Conference on Artificial Intelligence*, pp. 827-828. Tokyo, Japan; 1979).

[188] Rieger, C. and M. Grinberg: The causal representation and simulation of physical mechanics (Technical Report TR-495, Department of Computer Science, University of Maryland; 1976).

[189] Riesbeck, C. K.: Micro ELI (In R. C. Schank and C.K. Riesbeck (Eds.): *Inside Computer Understanding*, pp. 355-370. Lawrence Erlbaum: Hillsdale, NJ; 1981).

[190] Riesbeck, C. K. and C. E. Martin: Direct memory access parsing (Technical Report 354, Computer Science Department, Yale University; 1985).

[191] Rogers, B.: Probabilistic causality, explanation, and detection (*Synthèse*, **48**, pp. 201-223; 1981).

[192] Salmon, W.: *Statistical Explanation and Statistical Relevance*. (University of Pittsburgh Press: Pittsburgh, PA; 1971).

[193] Salmon, W.: *Scientific Explanation and the Causal Structure of the World* (Princeton University Press: Princeton, NJ; 1984).

[194] Sanford, A. J.: *Cognition and Cognitive Psychology* (Basic Books: New York, NY; 1985).

[195] Sanford, A. J. and S. C. Garrod: *Understanding Written Language: Explorations of Comprehension beyond the Sentence* (Wiley: New York; 1981).

[196] Schaffer, C.: In Kuhn's wake (Technical Report ML-TR-23, Department of Computer Sciences, Rutgers University, New Brunswick, NJ; 1988)

[197] Schank, R. C.: *Conceptual Information Processing* (North-Holland: Amsterdam, The Netherlands; 1975).

[198] Schank, R. C. and R. P. Abelson: *Scripts, Plans, Goals and Understanding — An Inquiry into Human Knowledge Structures* (Lawrence Erlbaum: Hillsdale, NJ; 1977).

[199] Schank, R. C., and C. K. Riesbeck (Eds.): *Inside Computer Understanding.* (Lawrence Erlbaum: Hillsdale, NJ; 1981).

[200] Schlimmer, J.: Conceptual learning from examples (*Proceedings of the Tenth International Joint Conference on Artificial Intelligence*, pp. 511-515. Milan, Italy; 1987).

[201] Shafer, G.: *A Mathematical Theory of Evidence* (Princeton University Press: Princeton, NJ; 1976).

[202] Shaver, K. G.: *The Attribution of Blame — Causality, Responsibility and Blameworthiness* (Springer-Verlag: New York, NY; 1988).

[203] Shaw, M. J.: Knowledge-based scheduling in flexible manufacturing systems — An integration of pattern-directed inference and heuristic search (*International Journal of Production Research*, **26,** pp. 821-844; 1988).

[204] Shibatani, M. : *A Linguistic Study of Causative Constructions* (Indiana University Linguistics Club: Bloomington, IN; 1975).

[205] Shortliffe, E. H. and B. G. Buchanan: A model of inexact reasoning in medicine (*Mathematical Biosciences*, **23**, pp. 351-379;1976).

[206] Shoham, Y.: Chronological ignorance — Time, nonmonotonicity, necessity, and causal theories (*Proceedings of the Sixth National Conference on Artificial Intelligence*, **1**, pp. 389-393. Philadelphia, PA; 1986).

[207] Shoham, Y.: *Reasoning about Change: Time and Causation from the Standpoint of Artificial Intelligence* (The MIT Press: Cambridge, MA; 1988).

[208] Simon, H. A.: *Models of Man* (Wiley: New York, NY; 1957).

[209] Slagle, J.: *A Heuristic Program That Solves Symbolic Integration Problems in Freshman Calculus — Symbolic Automatic Integrator, SAINT* (Ph.D. Dissertation Rept. 5G-0001, Lincoln Laboratory, Massachusetts Institute of Technology; 1961).

[210] Stanfill, C. and D. L. Waltz: Toward memory-based reasoning (*Communications of ACM,* **29**, pp. 1213-1228; 1986).

[211] Subbotin, M. M.: Computer applications and the construction of chains of reasoning (*Automatic Documentation and Mathematical Linguistics*, **20**, pp. 1-10; 1986).

[212] Suppes, P.: *A Probabilistic Theory of Causation* (North Holland: Amsterdam, The Netherlands; 1970).

[213] Taylor, R.: The metaphysics of causation (In E. Sosa (Ed.): *Causation and Conditionals*. Oxford University Press: Oxford, Great Britain; 1975).

[214] Tonge, F.: *A Heuristic Program for Assembly Line Balancing* (Prentice-Hall: Englewood Cliffs, NJ; 1961).

[215] Trabasso, T. and P. van den Broek: Causal thinking and the representation of narrative events (*Journal of Memory and Language*, **24**, pp. 612-630; 1985).

[216] Trabasso, T. and L. L. Sperry: Causal relatedness and importance of story events (*Journal of Memory and Language*, **24**, pp. 595-611; 1985).

[217] Turner, R.: *Logics for Artificial Intelligence* (Ellis Horwood: Chichester, Great Britain; 1984).

[218] Vendler, Z.: Agency and causation (In P. A. French, T. E. Uehling and H. K. Wettstein (Eds.): *Midwest Studies of Philosophy*, *Vol. IX — Causation and Causal Theories*, pp. 371-384. University of Minnesota Press: Minneapolis, MN; 1984).

[219] Waterman, D. A. and Hayes-Roth, F. (Eds.): *Pattern-Directed Inference Systems* (Academic Press: New York, NY; 1978).

[220] Waters, R. C.: The programmer's apprentice — A session with KBEmacs (*IEEE Transactions on Software Engineering*, **SE-11**, pp. 1-12; 1985).

[221] Weiss, S. and C. Kulikowski: Expert — A system for developing consultation modes (*Proceedings of the Sixth International Joint Conference on Artificial Intelligence*, pp. 942-947. Tokyo, Japan; 1979).

[222] Weiss, S. M., C. A. Kulikowski and S. Amarel: A model-based method for computer-aided medical decision-making (*Artificial Intelligence*, **11**, pp. 145-172; 1978).

[223] Whittlesea, B. W. A.: Preservation of specific experiences in the representation of general knowledge (*Journal of Experimental Psychology: Learning, Memory and Cognition*, **13**, pp. 3-17; 1987).

[224] Wilensky, R.: *Understanding Goal-Based Stories* (Ph.D. Thesis and Technical Report 140, Department of Computer Science, Yale University; 1978).

[225] Wilkins, D.: Knowledge base refinement using apprenticeship (*Proceedings of the Tenth International Joint Conference on Artificial Intelligence*, pp. 646-651. Los Angeles, CA; 1988).

[226] Wilks, Y.: What sort of taxonomy of causation do we need for language understanding? (*Cognitive Science*, **1**, pp. 235-264; 1977).

[227] Yager, R. R.: Linguistic representation of default values in frames (*IEEE Transactions on Systems, Man and Cybernetics*, **SMC-14**, pp. 630-633; 1984).

[228] Zadeh, L. A.: Fuzzy sets (*Information and Control*, **8**, pp. 338-353; 1965).

[229] Zadeh, L.: Fuzzy sets as a basis for a theory of possibility (*Fuzzy Sets and Systems* , **1**, pp. 3-28; 1978).

[230] Zadeh, L. A.: Syllogistic reasoning as a basis for combination of evidence in expert systems (*Proceedings of the Ninth International Joint Conference on Artificial Intelligence*, pp. 417-419. Los Angeles, CA; 1985).

[231] Zobrist, A.L. and F. R. Carlson, Jr.: An advice-taking chess computer (*Scientific American* , **228**, pp. 92-105; 1973).

Subject Index